D0457903

A SHEARWATER BOOK

Where Our Food Comes From

Where Our Food Comes From

Retracing Nikolay Vavilov's Quest to End Famine

Gary Paul Nabhan

Foreword by Ken Wilson

ISLANDPRESS / Shearwater Books
Washington • Covelo • London

A Shearwater Book
Published by Island Press

Library of Congress Cataloging-in-Publication Data.
Nabhan, Gary Paul.
Where our food comes from : retracing Nikolay Vavilov's quest to end famine / by Gary Paul Nabhan ; foreword by Ken Wilson.
p. cm.
Includes bibliographical references and index.
ISBN-13: 978-1-59726-399-3 (cloth : alk. paper)
ISBN-10: 1-59726-399-0 (cloth : alk. paper)
1. Centers of plant diversity. 2. Vavilov, N. I. (Nikolai Ivanovich), 1887–1943. 3. Food crops—Germplasm resources—Collection and preservation. I. Title.
QK46.5.D58N33 2009
581.6'32—dc22 2008013945

British Cataloguing-in-Publication data available.

Printed on recycled, acid-free paper 

Design by Brian Barth
Manufactured in the United States of America
10 9 8 7 6 5 4 3 2 1

Keywords: Famine, hunger, agriculture, farming, seed bank, biodiversity, cultural diversity, Russia, Stalin

For Laurie Smith Monti, who traveled with me retracing Vavilov's steps and who works daily to ensure that biodiversity, cultural diversity, and environmental health will be shared with future generations.

Nikolay Vavilov, from sketch on display at the Vavilov Institute (VIR).

Contents

Foreword

Food is back on the mainstream agenda. Some eight hundred million people can't afford the food they need and an even greater number—now some one billion people—are obese and suffering from unbalanced diets. Industrializing food production by making production directly dependent on fossil fuels was amazingly successful in the short run, especially if you didn't take into account the rest of the ecosystem, and whether food was about more than nutrients and profit, and didn't consider food's relationship to cultural identity, communing with the senses and with family, and delighting in the diversity and abundance of creation. Over the long run, which seems to be arriving at approximately *this moment*, such use of cheap energy to consume unlimited water and pour chemicals into soils has—as many anticipated—led to the specter that peak oil is now linked to peak water is now linked to peak soil and so to peak food. And all long before peak consumerism and peak population.

When food becomes just a commodity and interchangeable with energy we get the bio-fuels confusion. You get your SUV, linked to your army and linked to your wallet, achieving satisfaction before my children's stomachs, and those of our grandchildren. You get the subsidy that enables you to be paid to convert energy into food and back into energy at a net loss. And you get the predictable claims whenever a system reaches breaking point that what we need is just one more technological breakthrough to fix what's starting to fail—rather than rethinking a new system out of the old. If food is back on the mainstream agenda, this book is the inside and underside of why.

Where does our food come from? Much of humanity no longer knows how to put their hands into the soil, and instead specializes in curating, with microwaves, long-dead objects found in fridges.[1]

Rural people—let alone farmers—are treated as endangered species to be patronized at best. We no longer have a way to learn from the fact that indigenous tribe X managed to feed themselves sustainably in a desert for several thousand years with only a few crises. Where is our science, we ask, when we hear that the world has destroyed a quarter of the irrigable land, lost thousands of tasty varieties of crops and livestock, and allowed farming economies and rural landscapes to break down? Hearing from scientists that "things are complicated" and generally "not good," we join social movements and we despair of the powerful (while they tell us through their PR firms that they're joining us).

As usual, we think our times are unique. But as this fabulous book lays out, many of these same struggles were already under way in the first half of the twentieth century, when the roots of our current situation were sown. Therefore, to understand what we face in the twenty-first century we could do well to understand the science and politics of *Where Our Food Comes From*. One of the principal actors in this tale is a scientist few Americans have ever heard of; and most of those who have, have apparently misunderstood.

In the great Russian scientist Nikolay Vavilov (1887–1943), plant geneticists worldwide find an enduring hero, a founder of modern crop breeding, a man who died defending science in the face of Lysenko and political expediency as Soviet science rotted under Stalin and millions of peasants were silenced and starved. In this path-breaking book Gary Nabhan gently but persistently reminds us that in Vavilov we also find another scientific hero, and one almost unknown. Ironically this "other Vavilov" is a hero for environmental and social justice activists troubled by the unintended consequences of that same post-WWII crop breeding revolution that Vavilov's discoveries helped to usher in. These consequences included the spread of industrial farming and the "green revolution" that contributed to the destruction of diversity in crops and their wild relatives, the neglect of ten thousand years of indigenous farming traditions, and erosion of the benefits arising from successful coevolution between landscapes, cultures, and foodways. Thus increasing yields in fields went side by side with decreasing resilience and sustainability in agricultural landscapes, and with a loss of cultural heritage.

Vavilov's ability to create founding ideas on both sides of the current global food crisis debate certainly reflects the brilliance of this unrestrainable thinker. Yet it also reflects the fact that his prime piece of research equipment was a mule (in fact, a large number of them, able to follow any hill path or field line and arranged rather like a parallel processor, with the redundancy that allowed the occasional one to fall into a ravine without all his samples being lost). Until, and perhaps despite, the development of Google Earth, it has been hard to beat the mule as a tool for understanding human ecology. Dr. Nabhan's ability to lead us, at this same close-observing speed, in Vavilov's footsteps marks this book uniquely as a biography not only of Vavilov but also as a biography of the very agricultural systems that informed the birth of agricultural genetics.

Nabhan shows us too that Vavilov had another great methodological asset for understanding crop varieties and how they fared in the field: his willingness to learn local languages and embrace farmers as colleagues as he searched across five continents for the meaning of diversity. Thus armed with the mule and the capacity to learn from the custodians of the seeds, Vavilov became the first to recognize for science that there was a geographic coherence and pattern to those ancient agricultural civilizations of mountain tribal peoples who had domesticated the thousands of plant species upon which we still depend for our food. These became the "Vavilovian Centers of

Diversity," which remain, with some modifications, the foundation stone of agricultural genetics. It is a fact worth reflecting upon that almost nothing has been domesticated in modern times, and that science has actually failed to develop any new crops at all.[2]

Where our food comes from, then, is the creativity of long-past generations and the ways that peoples around the world differently encountered and imaginatively transformed biodiversity. Indeed, as Nabhan reveals, Vavilov seems to have been the first scientist to suggest that there were actually correlations between the cultural diversity in a landscape and the diversity of its agriculture and crop varieties. Vavilov understood that the patterns of genetic variation that a crop breeder might be interested in would not be random, but instead shaped by millennia of indigenous breeding in specific landscapes. It was therefore important to understand farmers' seed-selection efforts, to appreciate what they were seeking in diversity, and to reflect on the implications of the ecology of different farming systems. In short, hidden in Vavilov's initial insight and method are key ideas desperately needed if we are to sustain food production in a time of climate unpredictability, water scarcity, the spread of weeds, pests, and diseases, and the other woes inherent in our current global system.

In the course of this book Gary Nabhan takes us through the very same landscapes—intellectual and physical—that Vavilov traversed in those interwar years. He weaves together in his gentle lyrical style his own observations with an account of Vavilov's travels, drawing systematically on those documents, photos, and field notebooks that survived Vavilov's arrest and humiliation. In doing so, he allows us to see these landscapes in stereo vision—and with resultant delicious depth—through two sets of eyes some seventy to ninety years apart. Nabhan himself exemplifies many of the same skills with which Vavilov was so richly endowed: an ability to move lovingly at landscape speed through fields and communities, sampling, measuring, analyzing, and pondering; an ability to listen closely to people engaged with the land and indeed everyone with something interesting to say; and an ability to elaborate and illuminate general theory from an extraordinary grasp of the telling detail.

Retracing Vavilov's voyages of discovery to the mother lodes of agricultural diversity is a sobering process. We all know what's happening to the diversity of rainforests and coral reefs, but few realize that the situation is no less perilous for crop varieties. Numerous studies both local and global show how rapidly we are losing the genetic diversity of our crops and animal breeds, and the detailed knowledge of the meaning and maintenance of that diversity among our farmers, herders, food makers, and landscape managers. For example, the United Nations Food and Agriculture Organization has estimated that we have lost three quarters of the biodiversity of our crops over the last century and one animal breed per month over the last seven years.

In this context, renewed efforts are now under way to sustain samples of the world's crops in seed banks around the world—the very institutions that Vavilov first imagined and created to facilitate crop breeding—going so far as to create one in an arctic vault in Norway, a natural form of cold storage, in the event of local or global collapse.[3] And farmers and their allies everywhere are insisting upon maintaining these seed varieties in fields where the crops continue to coevolve with their pests and the environment, and where the knowledge of how to craft agricultural landscapes to embed this diversity has to be sustained over centuries and millennia because it cannot be redeveloped in any less time. All of these efforts to protect variety are particularly important in the "Vavilovian Centers of Diversity," where the greatest concentrations of biological and cultural coevolution have been located for nearly all the crop types that now circulate the globe. How many American beer drinkers would have guessed that it was Ethiopian farmers who provided the essential genes for rust-disease resistance for barley production in the United States? How many know that USAID and U.S. nongovernmental organizations (NGOs) are nevertheless still trying to persuade Ethiopians to abandon these traditional varieties? Thus when we walk into those fields with Nabhan and share with him the images of those farmers' hands and tools and the musical cultures that accompany and inspire their work, we desperately need to know whether the locals are succeeding in maintaining the diversity of their crops and the knowledge of how to sustain that diversity in the years to come.

Many factors are shaping these threats to agrobiodiversity, and most are connected to the rise and spread of industrial agriculture and food systems that remain stuck in the twentieth-century paradigm of achieving scale through homogenization and simplification. In agriculture, this means growing just a very few types of food in a very few ways in landscapes that have been extensively simplified and even made regularly shaped to enable machinery and inputs to be applied copiously without the inconvenience of taking nature's messy complexity into account. Such systems are also expanding because they can be more easily understood, managed, and replicated by the distant and powerful. The long-term consequences for these often apparently highly productive systems include low resilience (specialization against the realities of climate and other variability means farmers need bailing out of some crisis or another every few years), poor sustainability (short-term returns compromise long-term productivity), erosion of livelihoods and communities (labor, skills, and values are shed as monocultures take over), over-exploitation of water, and damage to other environmental services.

Tragically, diversity loss is not only being driven by economic interests. Across the world thousands of NGOs are joining ministries of agriculture and other groups in the endless distribution of what they call "improved" seeds and livestock, which on closer study turn out to often be poorly identified and rarely locally tested, usually already (or soon proven by study or experience) inferior to local types except in ideal conditions. Even when farmers riot in protest there seems little that is learned, because what drives this model of development is a cultural idea of "modernization," in which the outside object is fetishized and local people are deemed ignorant or passively needing their share of "the benefits of science." Development in this sense is characterized by ascending to the dominant culture, and by stripping one's culture and landscape of its uniqueness and making it an imitation of the dominant society.

In the face of all these assaults diversity fortunately survives much better than might be expected. As Nabhan himself found in areas of the former Soviet

Union, even after seventy years of collectivization followed by fifteen years of Western aid agency development, both of which vigorously tried to suppress local varieties, farmers had quietly (and secretly, if necessary) managed not only to maintain an amazing amount of diversity, but continued to breed new varieties that out-competed what the agencies brought to them. In fact, studies from around the world show that farmers often take advantage of introduced varieties to find genes that improve traditional varieties or to re-breed introduced varieties to *look* like the more tasty traditional ones so that they can trick people into buying them in the marketplace at higher prices.

The integration of food production and energy use highlights climate change. The use of fossil fuels to simplify, fertilize, and control ecosystems for crops and to produce crop-fed, methane-generating livestock makes industrial agriculture a leading cause of greenhouse gas emission, and yet in the face of perceived food shortages agriculture is the sector with the fewest plans for mitigation. Climate change projections suggest that major shifts in agricultural systems are almost inevitable, even if we quickly curb emissions, because of changes we have already set in motion in temperature and water supplies for crops, and/or for the weeds, pests, predators of the pests, and other elements of the impoverished ecosystems we still call farms. (The close detail of Vavilov's own observations in the Pamir Mountains enables Nabhan and his collaborators to show just how much change is already happening in the altitudes at which crops are grown.) After influencing global carbon dynamics for several millennia, initially by clearing forests and developing paddies or by discovering how to fix vast quantities of carbon in soils as "terra preta" (a black soil created over many centuries by incorporating carbon into poor soils in places like the Amazon), agriculture, in most parts of the world, is now going to suffer the consequences of its actions. We need to imagine and realize major adaptive change.

In the search for solutions to sustaining food production in the context of shifting environmental, demographic, and economic pressures, scientists and farmers are again divided by the Vavilov legacy: some point to gene management while others argue for an ecological approach informed by landscape, history, and culture. This argument tends to pit those who argue for "maximizing" models, in which simplification and specialization drive increased production, against those who argue for diversity and resilience in "adaptive management" approaches. These adaptive approaches seek to achieve food production at lower cost to long-term ecological health by understanding the complexities of agricultural landscapes through farmer knowledge and ecosystem science.

Maximizers take heart from the current revolution in the science of genetics (which Vavilov would doubtless have found fascinating), which will enable a whole suite of technical interventions to be made in the genomes of our food. This degree of intervention was hardly imaginable even a decade ago, and is far beyond the clunky genetically modified organisms (GMOs) that are currently so hotly contested. Meanwhile, the "adaptive management" advocates point for support to the other big contemporary revolution in science—namely, how grappling with the workings of "complex dynamic systems" helps us to rethink "productivity" questions (because they enable us to see that a gain in one part of the system may be a cost in another), and to understand the implications of uncertainty. This is the crucial "techno-garden" versus "adaptive mosaic" question as laid out in the Millennium Ecosystem Assessment scenarios that every world citizen should review, and which is central to the debate between the group of (sometimes green) technologists who think we can find ways to use adaptive artificial intelligence to control the complexity of Gaia and the cultural ecologist types who argue that the key to a successful future lies in learning from and adapting to Gaia's own rhythms. Thus the concept of "adaptive mosaic," which maintains its richness through encouraging places and cultures to continue to co-evolve with both local and global forces.

These debates on the future of agriculture and food tend to get over-simplified and stereotyped, often around flashpoints like GMO crops. This is unfortunate because crop breeding as conceived by Vavilov and others does not have to be about mono-cropping in energy-subsidized systems. Indeed, the very diversity of crops inherited by the modern age is the ongoing result of breeding by traditional farmers who have identified and recombined promising genetic traits for an astonishing variety of ends—as Vavilov discovered and countless detailed studies have since demonstrated. Thus it is not breeding as such that leads to the loss of diversity. The domestic dog, for example, is the most morphologically diverse vertebrate species and also the most highly selectively bred; this was achieved through breeding undertaken for changing and diverse purposes over long periods of time in a strikingly self-organizing way.

When we step back with Nabhan to view Vavilov's legacy we realize that it is not genetics and breeding that are the problem. Instead, it is certain kinds of breeding that result in centrally distributed identical materials, typically promulgated by state or private monopolies that are neither motivated nor able to deal with the long-term and unintended consequences of their actions and policies. There's a critical related question as well: whether this new genetic knowledge and technology will become public domain and readily available to farmers or the zealously guarded province of corporate interests. Basically, we need open-source crop breeding among farmer networks, just as we need it in software among tech networks.

Just as Vavilov described ninety years ago how farmers had developed and were distributing a wheat variety they had acquired through an in-law that could grow at a higher altitude than ever along the Afghan-Tajikistan border, so too are local farmers in many parts of the world today ever-ready to be leaders or partners in breeding experiments that respect their rights and knowledge. In other words, we will breed more and better new varieties when scientists and farmers collaborate in co-creating novel strains, rather than relying on geneticists to develop alone what they think will work best in that complex agroecological system we call a "field,"[4] that extraordinary chemical lab and factory referred to as a "kitchen,"[5] and for the delectation of culturally layered sensory primates otherwise known as, say, "fussy first child," "grandma," or the "allergic neighbor Bert." That greater effectiveness in plant breeding comes from allowing all knowledge to be applied to the problem we know from masses of experience (both positive and negative). The fact that other approaches still get the majority of funding is because of private interests, and sometimes because of the vanities or narrowness of training and perspective of the actors. However, this will change, just as it is changing in other domains of the knowledge economy. An approach to crop breeding involving a partnership with farmers will also recognize how farmers themselves are networked and how they distribute new varieties more quickly and more cheaply than can markets in commercial seed—seed which itself has been specially bred so as to require repurchase every year and which is protected by property laws that offend commonsense ideas about the work of nature and past generations in developing the genes they contain.

A powerful example of the inherent networked creativity that often resides in traditional selection and distribution systems is provided by the rapid spread in Colombia of a strain of coca resistant to the herbicide glyphosate. Glyphosate, patented as Roundup by Monsanto, is one of the most widely sold agro-chemicals of all time. It is the basis of the dominant GMO product, Roundup Ready soy (and now other crops), genetically engineered to resist the herbicide applied to kill the weeds around it. The U.S. government has been aerially spraying coca fields in Colombia with glyphosate as part of its "war on drugs," requiring farmers to relocate production to national parks or to develop their own "Roundup Ready" resistant varieties. When coca resistance to the herbicide did emerge it was assumed that the drug lords must have funded the pirating of GMO-coca in the lab, but investigations showed no evidence of tampering with the coca genome:

> [T]he implication is that the farmers' decentralized system of disseminating coca cuttings has been amazingly effective—more so than genetic engineering could hope to be. When one plant somewhere in the country demonstrated tolerance to glyphosate, cuttings were made and passed on to dealers and farmers, who could sell them quickly to farmers hoping to withstand the spraying. The best of the next generation was once again used for cuttings and distributed.
>
> By spraying so much territory, the US significantly increased the odds of generating beneficial mutations. There are numerous species of coca, further

> increasing the diversity of possible mutations. And in the Amazonian region, nature is particularly adaptive and resilient.

As Joshua Davis thus concluded in his delightful article in the renowned pro-technology magazine *Wired*: "In this war [on drugs], it's hard to beat technology developed 10,000 years ago."[6]

Interestingly enough, both the genetic-lab approach and the agro-ecosystem approach to developing new crop varieties are pointing us back toward the "wild relatives" that so fascinated Vavilov. These are the wild species from which crops were originally domesticated many thousands of years ago, and from which early farmers extracted characteristics that improved their productivity, responsiveness to management, harvestability, and qualities as foods. To create our crops these farmer-breeders secured dramatic transformations in many species, and in the process discarded significant chunks of genetic diversity (intentionally and unintentionally) even as they took advantage over the millennia of new mutations arising in their crops that might not have persisted or been mixed with the right other traits to thrive in the wild. Contemporary geneticists and plant breeders are currently much excited by the idea that we can now return to re-gather that diversity from the wild relatives to enable the breeding of crops to deal with such new and old challenges as drought tolerance and resistance to disease. Efforts of this kind are indeed already showing results in potatoes and in wheat.

These wild relatives are also understood as crucially important in indigenous farming traditions. In each of the regions Vavilov visited where the Christensen Fund works today, for example, farmers are still actively encouraging gene flow between their domesticated varieties and their wild progenitors for similar if less genetically precise aims. We work with farmers in Ethiopia who are finding and bringing into cultivation new strains of wild crops like coffee and enset, and with Kyrgyz Republic farmers who know how to mix wild apricots into their orchards because this improves pollination. The future of these wild relatives continues to be threatened by the spread of industrial agriculture and all the other undignified consequences of modern life, and we would do well (if we want to have a long and fruitful relationship with the plants we've brought into our homes) not to despise the families of these crops we've made our brides.

Ours is a millennial generation that typically expects everything to change and the past to be irrelevant. Some of us believe that given what is going on in the world now, we should expect an ending, perhaps even rapture, at any moment, or at the very least a monumental decline and fall. Others believe that we have generated so much knowledge and technology in the last few decades that we either have already or will soon surpass the need to draw for guidance on any of the last ten thousand or five million years of human experience of living on this planet. Experience to date suggests that an ending to everything anytime soon is rather unlikely (just more misery, even more unequally distributed and more graphically presented in the media). In regard to the second view, and largely because of the Web, it is clear that we are entering a new kind of knowledge society and economy; yet it is also clear that we are still pretty much the same savannah primate we were until only the other day when fortune or misfortune found us grubbing around in what was to become one of the first fields. The only really significant genetic changes in humans since then are adaptations to our various foodstuffs, as described in Nabhan's earlier book *Why Some Like It Hot*, and to the diseases brought on by living in settled larger populations that are themselves increasingly networked.

Faced with uncertainty about where we are headed, it doesn't seem sensible to cast aside knowledge of where we have come from. Human beings are engaged in fundamentally changing systems long before we understand them, and we take change to the global scale long before we have more than the vaguest knowledge of the consequences—yet still we sneer at the custodians of those historically informed farming systems and landscapes when they protest or offer words of counsel. Vavilov would have been aghast at where his discoveries have taken us, though doubtless in a rather dignified way. Since it is the mission of The Christensen Fund to back these traditional stewards and guardians of the world's cultural and biological diversity and help throw light on their contributions to the planet's future, we are delighted to have assisted Dr. Nabhan to undertake these journeys and bring these stories and values to wider publics.

Gary Nabhan's book takes us back to the master and invigorates us with "science" as the practice of inquiry and discovery (rather than the top-down delivery of judgment and technology). He illuminates agriculture as an extraordinary process where culture, technology, and biology, and much else besides, are applied to landscapes with fascinating consequences that need to be thought about. Agriculture as practiced locally in places like the centers of diversity Nabhan visited is something he also shows us is beautiful; it moves the palate as much as the eye and can inspire us to join movements like Slow Food, and to break the soil for a seed with a song, whether we be in a traditional terraced field in Mexico or an urban community garden in a mega-city.

K. B. Wilson
Executive Director, The Christensen Fund

Notes

1. As Carlo Petrini, the founder of Slow Food globally, has pointed out so eloquently.
2. The exception that proves the rule is perhaps two species of macadamia nut that were domesticated in Australia in the 1890s. However, this involved white settlers merely planting trees that had been harvested wild for millennia by Aboriginal Australians, and subsequent breeding has done little to transform the plant except by making the shell easier to crack through breeding in blemishes, an advance achieved not by a plant geneticist but by a retired actor/estate agent/stockbroker turned Florida gardener, Mr. Morris Arkin. See http://en.wikipedia.org/wiki/Macadamia and http://en.wikipedia.org/wiki/Morris_Arkin.
3. Striking, however, is that nobody from Zimbabwe traveled to Norway to obtain seed to re-establish local crop production when the country's long-successful hybrid maize seed production system eventually collapsed in the current crisis. Instead, they visited that more accessible hub of heritage crop types, and one attached to distributive social networks rather than visa bureaucracies, the veritable local *ambuya* (grandmothers), who became the sources for the seed varieties that enabled the remarkable resurgence in sorghum and millet production in recent years that has fed the poor.
4. Miguel Altieri, *Agroecology: The Science of Sustainable Agriculture* (Boulder, CO: Westview, 1995).
5. Harold McGee, *On Food and Cooking: The Science and Lore of the Kitchen* (New York: Scribner, 2004).
6. See ww.wired.com/wired/archive/12.11/columbia.html?pg=5&topic=columbia&topic_set=.

Map of places Vavilov and Nabhan visited

ARCTIC OCEAN
ARCTIC OCEAN
RUSSIA
UNITED KINGDOM
NETHERLANDS
POLAND
BELARUS
GERMANY
UKRAINE
FRANCE
KAZAKHSTAN
CROATIA
BULGARIA
SPAIN
ITALY
KYRGYZSTAN
UZBEKISTAN
NORTH KOREA
GREECE
TAJIKISTAN
CHINA
JAPAN
SOUTH KOREA
CYPRUS
SYRIA
TUNISIA
LEBANON
ISRAEL
IRAN
AFGHANISTAN
JORDAN
ALGERIA
EGYPT
PACIFIC OCEAN
ERITREA
ETHIOPIA
INDIAN OCEAN

CHAPTER ONE

The Art Museum and the Seed Bank

During the White Nights of 1941—around the time of the summer solstice, when the twilight lingers beautifully and indefinitely in the skies of the northernmost latitudes—Hitler's forces first crossed from Poland into the Soviet Union, with their sights set on taking Leningrad. Soviet military intelligence was well aware of the seemingly endless caravans of German and Finnish troops, tanks, and artillery that were on the move, and surmised that those forces could converge on the city by summer's end. Stalin and his generals feared that if the German and Finnish forces were to take control of Leningrad—old Saint Petersburg—they would deal economic, strategic, and symbolic blows to the Soviets and their allies, for more than any other in the nation, that city was home to considerable monetary as well as artistic wealth.

On July 15, 1941, Stalin authorized an emergency evacuation of what the Soviets considered to be the city's most priceless treasures, those they believed the Nazis sought to confiscate and control for their own purposes. The rest of the world held its breath while the fate of those treasures was being decided by the head-on clash between two of the greatest armies ever assembled on the planet.

Most Western intellectuals were particularly concerned with the safety of the extraordinary art collections held at the Hermitage, one of the world's oldest and largest museums of human history and culture. Well over two million paintings, sculptures, coins, jewelry, and artifacts were housed there in the six hundred rooms of the Winter Palace, built by the czars in the heart of Leningrad.

While Stalin had been deliberating how to authorize an evacuation without causing panic in the population or admitting his own vulnerability, the keepers of those cultural treasures in Leningrad had already taken action. It took but two days after Hitler's invasion of the country for the Hermitage director, Iosif Orbeli, to initiate a plan for emptying Russia's greatest art museum. He recruited not only his curators, but also hundreds of artists, historians, students, and laborers. With the utmost urgency, they would have to take roughly a million paintings out of their frames, label them and roll them up or pin them down in boxes, and cushion them in packing material so that they could be hidden away. In a mere six days, more than a million and a half works of art were readied for secret storage in vaults hidden in the Hermitage basement, in a nearby cathedral, and in the hinterlands of the Russian steppe. Before dawn on the morning of July 6, 1941, a half million paintings, drawings, frescoes, artifacts, gems, vessels, and ornaments from the Hermitage were boarded onto the first train leaving Leningrad, headed to sanctuaries in a locality known only to a few Soviet officials. On July 10, another seven hundred thousand masterpieces, filling fifty-three Pullman cars, were sent toward the village of Sverdlovsk some 2,500 kilometers away. There they would spend the next three years cloistered in a Catholic church, sequestered in an art gallery, or sentenced to the death-tainted basement of the Ipiatev Mansion, where the family of Czar Nicholas II had been shot almost three decades before. The best and brightest of the Hermitage's conservation staff were dispatched to stand watch over the collections in Sverdlovsk, to protect them from fires, looters, and other potential dangers.

The interpreters and guides at the Hermitage today love to detail how successful those efforts were in safeguarding some of the greatest masterpieces of the Greek, Roman, Medieval, and Renaissance eras. They seldom if ever mention, however, another priceless world-class collection of our shared heritage that lay just a few blocks away, unheralded, on Saint Isaac's Square.

That second treasure trove harbored—and harbors still—more than 380,000 living, breathing samples of seeds, roots, and fruits of some 2,500 species of food crops that had been collected by Russia's world-class cadre of plant explorers who had worked for the Bureau of Applied Botany since 1894. Those seeds came in all colors, sizes, and shapes, some dull-coated while others glistened like jewels, as if hinting at a more priceless bounty of diversity still out in the fields of peasant farmers around the world. The tubers, roots, and bulbs came in all sorts of textures, from knobby and gnarled to as smooth and burnished as a clay pot shaped on a wheel, glazed, then fired in a kiln. The myriad fruits exuded nearly every fragrance imaginable to a perfume chemist—musky, fermented, citric, and floral. The fruits and nuts came in all kinds of arrangements, from cascading clusters of berries to the geometric wonders of pineapples and pine cones. Most of them were not only good to gaze at, like the art in the Hermitage, but exceedingly good to eat.

That treasure had myriad potential uses: The seeds could be multiplied and distributed to farmers, who could grow them to feed their families; selections of seeds could be used by plant breeders to improve the disease or pest resistance of more vulnerable varieties whose susceptibility was leading to famines or food shortages; some deeply rooted varieties were useful for soil erosion control and for the restoration of damaged landscapes; still others were key to unlocking the stories of where our food originally came from, helping us to elucidate the origins of agriculture and the earliest domestication of plants on several continents. Some seeds had remarkable stories associated with them, and all had genetic histories embedded within their seed coats. Most of the seeds were priceless, in the sense that they could not easily be re-collected or replaced, for the agricultural landscapes from which they had been derived had changed dramatically over the previous century. They represented dynamic populations of plants that shifted and evolved through place and time—if they were lucky enough to avoid political and physical upheavals—and, for that reason, were all the more irreplaceable.

Yet few Russian residents passing by the seed bank hidden within the bowels of a stodgy building on Saint Isaac's Square ever fathomed its paramount significance to human survival, let alone its uniqueness as a living record of some of the greatest achievements made by the diverse cultures of this planet. In 1941, even fewer of the artists, intellectuals, politicians, and bureaucrats distraught over the impending fate of the Hermitage could have imagined that the German troops engaged in Operation Northern Light were just as eager to control this genetic repository of seeds as they were to capture and sell off the artistic treasures housed in the Winter Palace.

Despite the damage done to Leningrad during the *Blokada* that began that September—a siege that lasted for nine hundred days and eliminated 1.5 million human lives from that bleak landscape—the building on Saint Isaac's Square that housed that priceless bank of the world's seeds miraculously survived. It has remained on the square to this day, harboring both seeds and scientists associated with what's known as the N. I. Vavilov All-Russian Scientific Research Institute of Plant Industry. The institute is nicknamed VIR by the relatively few Russians alive today who recognize its vital place in history and honor the memory of its charismatic founder, Nikolay Ivanovich Vavilov (1887–1943). Vavilov's legacy is more than just the seeds he collected from around the world, for what he most valued were the seeds that remained in a peasant's field, adapting and changing, along with the traditional knowledge of where, when, and how to plant them.

My friends at VIR cannot tell the story of this seed legacy without tears welling up, for their story ultimately leads to the fate of the seed bank that sits below their offices today. Although I had met a director of VIR in Rome

Vavilov in the field, Ethiopia, 1926.

in the 1980s while working as a consultant to the United Nations Food and Agriculture Organization (FAO), I didn't get to VIR itself until the spring of 2006. I went to Saint Petersburg then with an old friend, Kent Whealy, cofounder of the Seed Savers Exchange and recipient of the Vavilov Medal in honor of work he had done to conserve heirloom seed stocks and bring them back to our tables. Over the years, Kent and I had heard from Russian friends something of what had occurred at VIR during the darkest hours of the Siege of Leningrad. But we both wanted to hear the stories told in the place where they occurred—the heart of Saint Petersburg—by the very people who best knew Vavilov and those he had entrusted with keeping his seeds alive.

As our colleagues reminded us, in 1941, none of the support offered to the Hermitage staff was offered to those in charge of Vavilov's seed bank and the farms in the surrounding countryside—known as plant introduction stations—where the seeds were periodically grown out and replenished. Yet, from what Vavilov's staff knew of the strong German interest in eugenics, they could not imagine that the Nazi bureaucracy did not realize the importance of their genetic repository. As German and Finnish forces drove toward the city, the VIR staff feared that the Nazis would confiscate whatever seeds were available in the plant introduction stations associated with VIR's mission. The staff was at least able to hide some of the Saint Isaac's seeds at an experimental farm adjacent to Catherine the Great's palace in the suburb of Pushkin, just outside of Leningrad. But no staff was granted safe passage away from the fray. VIR's employees were to remain at their desks, continuing to do the work they had been assigned, as if neither war nor any other pressure was plaguing them.

By the end of the first autumn of the Blokada, Leningrad had been fully surrounded, and no food or fuel could reach the millions of Russians remaining in the city. While artillery fire escalated, food supplies dwindled to a thirty-day supply and were strictly rationed—down to 125 grams per person, or about a quarter pound of bread daily. Then the harshest and bitterest of winters set in, leaving the stranded masses with no heating oil or coal, little firewood, limited electricity, and, in most homes, no running water. Once grain and sugar supplies were depleted, families were given rations of mutton guts, malt flour, cellulose, and calf skins; both their health and their hope began to deteriorate.

By February of 1942, at least two hundred thousand people had died from starvation in Greater Leningrad or from the illnesses that pounced on the crippled immunity of the hungry. Despite those losses and their own lack of safety, many in Leningrad tried to continue their normal work, taking one day at a time. Those who had volunteered for the evacuation of the Hermitage could at least feel satisfied that they had done all that was possible to protect their city's most enduring works of art so that they might be enjoyed by future generations.

Much of that other great collection remained in grave danger, however. The extraordinary bank of living seeds that Vavilov had built and nurtured over the previous quarter century had been left exceedingly vulnerable. Reports had come in that the seeds left in the plant introduction stations in the Ukraine and Crimea had already been seized by the Germans; it was later learned that Heinz Brücher, a German geneticist, had sequestered them away at the Grannagh Castle in Austria. At the same time, even the portion of VIR's holdings that had been taken to the experimental station at Pushkin stood in harm's way. Pushkin was being shelled regularly, and the "Road of Life" leading beyond the city limits across the ice of Lake Lagoda was under such attack that it was renamed the "Road of Death."

In a daring move, the caretakers of the seeds loaded the portion of the collections held in Pushkin onto twenty trucks, whose drivers managed to pass through the German lines pretending to be peasants delivering grain to other German troops. That convoy of seeds eventually arrived, undetected, at the University of Tartu Experimental Station in Estonia in the summer of 1942. Those seeds thus fortuitously escaped the battle, but they could not escape the war. In the fall of 1944, the German army seized them in Estonia and began to pirate them off to Lithuania.

Unbeknownst to the VIR staff remaining in Leningrad, the life of their mentor Vavilov was then in as much peril as the seeds he had collected. For reasons I will later elaborate, Russia's greatest scientist had been taken as a political prisoner—not by the Nazis but by his own government—and was kept from public view while the Soviet government continued to issue press releases that he was simply helping Stalin and Soviet biologist Trofim Lysenko with a new strategy to feed the people. Although none of his coworkers had heard from him or of him since his departure for an "important meeting in Moscow" in the summer of 1940, they remained steadfast in their efforts to safeguard the seed bank. Even while starving, they demonstrated as much dedication to their mission as did as their counterparts in Sverdlovsk.

The only difference—a critical one—was that Stalin supported the evacuation of the Hermitage but considered the seed bank to be a costly indulgence of "bourgeois science." Although the Nazis could see the value for future plant breeding of controlling the world's largest seed bank, Stalin's Soviet cronies looked at the state support of the seed bank as a tremendous financial burden that had not offered much in return. Stalin had jailed Vavilov and dozens of other scientists for being elitists and traitors whose research had paid few dividends to the Russian peasantry or to the state itself.

The staff remaining in VIR's building on the square continued to work, with virtually no government support. They feared that the hungry masses lurking in the streets outside might attempt to break into their stores and consume the bags of wheat, barley, beans, and peas that the staff had hoped would provide the stock to feed future generations. So they barricaded themselves inside the stout walls of their building on Saint Isaac's Square and stood watch over the living collections that they hoped would help Russia and the rest of the world recover, should the war ever end. The workers, led by Abraham Kameraz and Olga Voskresenkia, divided the most valuable of the four hundred thousand seed collections into duplicate samples and put them into boxes for hiding at different locations. Kameraz personally convinced a small tactical unit of the Red Army how important it was to everyone's future to remove a set of these seeds to another building off Saint Isaac's Square.

The ensuing tragedy has often been recounted to scientists visiting VIR, but hearing it in person still sickened and silenced Kent and me. The scientists and curators locked themselves into the dank, unheated building, guarding the other set of seeds as well as all of their potatoes in the dark, damp conditions of the near-freezing basement. Numb with cold and stricken with hunger, the staff took shifts caretaking the seeds around the clock. Nine of Vavilov's most dedicated coworkers slowly starved to death or died of disease rather than eat the seeds that were under their care. They were not alone. Over seven hundred thousand citizens of Leningrad had died from hunger by the spring of 1944, when the siege finally ended.

Perhaps it is fortunate that the starving seed guardians never learned how close their building was to being seized by the Nazis, for such news might have broken their spirits and weakened their tenacity. Unbeknownst to any outsiders, the inner circle of Nazi strategists had early on targeted the Russian seed bank as being a far more important collection to capture than that of the art they believed to be still within the Hermitage. As soon as Hitler set his mind on invading Russia in 1941, he established a special tactical unit of the S.S.—the *Russland-Sammelcommando*—to take control of the seed bank and retrieve its living riches for future use by the Third Reich.

That Hitler had a prevailing interest in the genetic research of the Russians should come as no surprise. Hitler based much of his racist philosophies on the pseudoscience of eugenics, which argued not only for the selective breeding of humans for racial improvement, but also for the advancement of agriculture through highly selected seeds; his agricultural programs were managed by scientists influenced by social Darwinism and eugenics. Some have claimed that Hitler became a strict vegetarian and raw-foodist, as well as an adherent to some rather bizarre philosophies of dietary purification that complemented his notions of racial purification. Hitler or his scientific advisors may thus have envisioned ways that the diversity of seeds stored in Leningrad could ultimately serve those purposes.

Assuming that he would soon take control of Leningrad, Hitler had planned to make his victory speech from the balcony of the Hotel Astoria on Saint Isaac's Square, from which he would literally look out across the street to where the treasure of seeds lay before him, barely a hundred feet away. He even had invitations to the victory celebration printed up, with the address of Hotel Astoria prominently featured. The seed guardians never knew that Hitler planned to celebrate his conquest on their very doorstep.

While that was set to unfold in Leningrad, another tragedy was playing out in the little town of Saratov, some nine hundred kilometers away, on the Volga. That was where, in 1918, Vavilov had been awarded the title of professor in agricultural sciences at the rather precocious age of thirty-one. In 1941, Vavilov found himself back in that beloved town in the heart of the wheat-growing country where his career as a seed conservationist had begun to skyrocket; this time, however, he was a political prisoner, not a professor.

Vavilov—the only man on earth who had collected seeds of food crops on all five continents, the explorer who had organized 115 research expeditions through some sixty-four countries to find novel ways that humanity could feed itself—was himself dying of hunger. From the spring of 1942 until his death in January of 1943, having been fed nothing but a raw mash of flour and frozen cabbage, Vavilov was undernourished and emaciated, with little subcutaneous fat left on his skeleton. He suffered from chronic diarrhea, an itchy edema had broken out on his legs, and his muscle tissue had wasted away to the degree that he would soon be diagnosed with dystrophy. Adding insult to injury, the KGB was trying to break him down mentally, interrogating him for as much as fourteen hours a day. They were intent on making him confess that he had squandered the Soviet Union's financial resources to build his own empire of a hundred-some field stations to conserve and evaluate the world's plant diversity; in their minds, the seed empire was an extravagant diversion from the task of immediately feeding the masses. A famine in the mid-1930s caused by forced collectivization and confiscation of grain by the Soviet government had killed at least five million people, and now, with trade routes disrupted within and beyond the Soviet Republics, the Russian populace had less food security than ever before.

Vavilov's colleagues could never have fully fathomed the perils that both their seeds and their former leader had been facing around that time. When the summer of 1942 came, they planted cabbages and seed potatoes in the churchyard of Saint Isaac's Cathedral and in the fields surrounding the czar's former palace at Pushkin. The previous winter, they had found enough fuelwood to heat the Pushkin storehouse where some of the tubers lay dormant, hoping to keep the seed potatoes from perishing. Unlike other years, they had to stand guard over the potato plants twenty-four hours a day—sometimes while artillery fire sailed over their heads—and attempt to replenish the tuber stock to keep it viable for larger plantings in the future. Not only did they have to discourage their fellow Russians, who hungrily eyed each row of potatoes, but they also had to kill hundreds of Norway rats that invaded their garden.

Years later, the Russian writer Genady Golubev interviewed Vadim Lekhnovich, one of those who had helped dig the frozen earth, guard the sprouts, and watch over the garden of edible delights that spring. Was it hard, he was asked, not to help himself to some of those selections when he had already been starving for many months?

"It was hard to walk," he replied. "It was unbearably hard to get up every morning, to move your hands and feet. . . . But it was not in the least difficult to refrain from eating up the collection. For it was *impossible* [to think of] eating it up. For what was involved was the cause of your life, the cause of your comrades' lives."

After the current VIR staff had recounted what they knew of this history to Kent and me, one of them—Dr. Sergey Alexanian—told us that he'd like to offer us one last insight regarding his predecessors who had worked for

VIR during the siege. An Armenian of slight build, but with a powerful command of words and gesture, Sergey was conversant with both political history and agricultural sciences in ways that made him the ideal guide. He led us to a display of aging black-and-white photo portraits of former VIR staff on the stairway just outside his office. He wanted to explain how the sudden explosion of rats invading the potato gardens of Leningrad and Pushkin had occurred during the siege.

Nikolay Dzubenko, director of the Vavilov Institute, with Kent Whealy and Sergey Alexanian, April 2006.

"You see, the only meat available for people to eat by the summer of 1942 was that of the cats remaining in Leningrad. Without cats to control the rodent population, rats were out in the streets and yards night and day, digging up anything that might be edible. Do you see the picture of this woman here? She was in charge of the potato collection, and she died while protecting them from the rats. . . .

"And it was these men and women," Sergei said quietly, as he pointed to several other photos one by one, "who died while standing watch over the seeds."

Among this group, Alexander Stchukin died at his writing table, holding in his hand a packet of his most prized peanuts that he had hoped to send off for a grow-out. The custodian of Vavilov's many oat collections, Liliya

Rodina, died of starvation, as did Dimitry Ivanov, who as his own life failed, stowed away thousands of packets of rice that he had held so dear. There were others as well—Steheglov, Kovalesky, Leonjevsky, Malygina, Korzun—some who perished by starving, some riddled by sickness, others by shrapnel. Wolf, the herbarium curator, was hit by a missile shell fragment, and bled to death. Gleiber, the archivist of Vavilov's field notes, died in the midst of those papers rather than leave his post vulnerable to the infidels.

Kent and I stood next to Sergey, filled with sorrow. Several minutes passed before any of us could say a single word to one another, the presence of the Russians' sacrifice so palpable within those walls. The seeds had survived, but many of their proponents had not.

That same day, Sergey showed us row after row of the tin boxes where Vavilov once stored the seeds he had collected, as well as the gigantic vats of vaporous liquid nitrogen in which the progeny of those seeds remain frozen but viable. Knowing in advance that I was coming from the deserts of North America, where Vavilov had trod some three-quarters of a century before, Sergey had asked the herbarium curator to lay out the very plant specimens that Vavilov had brought back from there to the Soviet Union. Even in their desiccated, two-dimensional positions, plastered down and glued onto heavy sheets of herbarium stock, the plants felt like old friends: a thorny pad of prickly pear cactus, and a branch of guayule, a rubber-bearing shrub. I presented Sergey with photos taken of Vavilov by Homer Shantz, the plant geographer who had hosted the Russian scientist on his trips through the American deserts, and who had preceded me at the University of Arizona. While working on my master's degree in plant sciences at that university, I had stumbled upon a dozen images of the Russian scientist that had never been seen in his home institution. Now they could be integrated into VIR's archives with the thousands of field photos that Vavilov himself had taken.

Only after Kent Whealy and I had said good-bye and were on our way to see the fabulous art collections at the Hermitage, did I think of the question I wished I had asked Sergey and the other scientists: How was it that the art collections at the Hermitage could be so clearly seen as an important element in the common heritage of humankind, but an equally large and representative collection of seeds—of the very food we require for our physical survival—has been so blatantly undervalued by society at large?

Sergey had sidestepped such issues with us, but during his own tenure in the Vavilov Institute, its staff had been cut down to a fourth of what it had been in the years just prior to perestroika and a fraction of what it had been during Vavilov's tenure. Alexanian and the other fine scientists who are the current stewards of some 380,000 seed samples held by VIR are provided with resources for their work that are in no way commensurate with their talents as professionals or with the importance of their endeavor. That endeavor, simply stated, is to maintain the supply-and-delivery stream of genetic resources essential to feeding present and future generations, offering us a critical modicum of food security in the face of global climate change, new hyper-virulent pests and diseases, declining freshwater supplies, and the potential for wars and acts of terrorism to disrupt the transportation routes on which our food supplies flow.

Although most who work to conserve the biodiversity and vitality of our food resources in the United States and other developed countries are certainly paid better than their Russian counterparts, they still lack many of the essential resources needed to fully accomplish their tasks. In short, Russia is not the only nation that currently fails to make an adequate investment in the conservation of agricultural biodiversity in gene banks and as part of on-farm conservation. And now, to make matters worse, some decision makers have made the erroneous assumption that biotechnologies can "develop" in vitro all the genetic variation that will ever be needed to protect our food crops from disease, climate change, and other stresses, thereby making efforts to conserve biodiversity *seed by seed* somehow obsolete. More and more of the funding for biological sciences has shifted away from genetic conservation and evaluation and toward investments in biotechnologies, as if they were the ultimate panaceas.

The current paucity of funds to adequately conserve the biodiversity held in seed banks, botanical gardens, and experimental farms is only part of the problem, for the bulk of conservation work must be done on farms and in orchards in the regions to which the seeds are adapted. The seeds remaining in the very fields from which Vavilov gleaned them are at risk, as is the traditional knowledge among the farmers who know best how to cultivate them. Most of us living today hardly know where our foods come from. At best, we are dimly aware of the geographic and cultural origins of the crop genetic resources that form the living foundations of our food supply. We seem to believe that as long as we wish to eat, those resources will be invariably provided to the seed curators, plant breeders, nurserymen, and farmers who make our agricultural supply-and-delivery chain function. But as the seed keepers in Leningrad realized in 1941, we are in a race against time to ensure that the remaining seed varieties on this earth are not extinguished like so many candles in a sudden gust. No biotechnology can "invent" or replace the genetic variability already present in the diverse seeds found in the fields of local farmers scattered around the world; we have barely begun to classify those seeds on morphological grounds, let alone understand their genetic relationships and potential uses. Whether or not biotechnologies will be used in developing new seed strains, those locally adapted seed varieties—which continue to be dynamically bred and selected by peasant farmers as they have for millennia—will remain the primary wellspring of—or "gene pool" for—all future crop improvement efforts.

Vavilov and his American friend Harry Harlan were among the first scientists to notice that traditional seed stocks were indeed blinking out; they recognized early on that agricultural modernization was driving into extinction some of the locally adapted varieties they had collected on their earliest expeditions. Coming back into the same agricultural regions of Asia two decades after his initial visits to particular fields and orchards around 1916, Vavilov was shocked to learn that the same seeds could no longer be found there. Such shifts particularly disturbed scientists such as Vavilov and Harlan, for they recognized those regions as the historic centers of origin of certain domesticated cereals, where certain varieties and their ancestors had been sown by an unbroken chain of farmers since the beginnings of agriculture.

In correspondence with one another, Vavilov and Harlan were among the first to articulate the concept of *loss of agricultural biodiversity* through the process now known as *genetic erosion*—the gradual and irrevocable diminishment of the gene pool from which new varieties would otherwise emerge.

During World War II, Hitler's geneticists may not have been aware that the planet's crop genetic resources were facing declines in diversity in some regions, but within another two decades, just about every scientific plant explorer had reported declines in food crop diversity in the regions where they worked. Like the more recent global declines in amphibians, at first the scattered reports seemed merely anecdotal and did not foster much alarm; by 1970, however, the consistency among many reports made it clear that such losses were pervasive, reaching nearly every agricultural region on the face of the earth.

Today, scientists take as a given what Vavilov first articulated, a message that ultimately cost him his life: Agricultural biodiversity is the cornerstone for building greater food security for humankind; without it, our food system will be crippled by pestilence and plague, drought and flood, global warming, and the economic or environmental side effects of globalization. Although Vavilov himself never used the particular term *agricultural biodiversity*, I think he would have embraced the following definition of it, one that I had the good fortune to help forge for current discussions at the FAO:

> Agricultural biological diversity is embedded in every bite of food we eat, and in every field, orchard, garden, ranch and fish pond that provide us with sustenance, and with natural values not yet fully recognized. It includes the cornucopia of crop seeds and livestock breeds that have been largely domesticated by indigenous stewards to meet their nutritional and cultural needs, as well as the many wild species that interact with them in food-producing habitats. Such domesticated resources cannot be divorced from their caretakers. These caretakers have also cultivated traditional knowledge about how to grow and process foods; such local and indigenous knowledge—just like the seeds it has shaped—is the legacy of countless generations of farming, herding, and gardening cultures.

What delighted Vavilov were the patterns of relationships that this diversity formed as it spread across the face of the earth: the gradients in the length of beards on the spikes of barley as they ranged from Turkey to Ethiopia or from the floodplain fields along the Silk Road to the high plateaus perched above them in the Hindu Kush. He relished the various names local farmers gave to the same group of beans, for he could use them as clues in tracing their origins and dispersals. He was fascinated by the shapes, sizes, and tastes of all the wild apples found along a single mountain range in Kazakhstan, for they offered him a fresh view of the location where the original domestication of these fruits may have occurred.

Many historians of science have assumed that Vavilov's greatest contributions to agricultural botany and conservation were the world collection of seeds, fruits, and tubers in Saint Petersburg and the notion that efforts to fit crops to their agricultural environments would do well to use the diversity of land races that indigenous farmers have already adapted to similar conditions. Seeds from those vast gene pools, whether put directly in the soil or used in plant breeding, are our best means of dealing with pests, droughts, diseases, soil nutrient deficiencies, salinity, and short growing seasons. Others might point to Vavilov's two-hundred-odd published journal articles and books that documented in a way no one had before the astonishing genetic variability among the world's major crops. There are, however, two other important elements of his legacy that have not been well recognized or fully understood.

Intellectually, Vavilov's greatest contribution to science may have been his articulation in a 1926 publication of the concept of *centers of diversity*. Although he first articulated them as centers of origin for cultivated plants, he explicitly chose areas where genetic variation within the gene pools of domesticated crops and their wild relatives was high for both. Contradicting most archaeologists of his time, he argued that the cradles of agricultural civilization were *not* nestled in the valleys or broad floodplains of great rivers, but were found in mountainous regions. Vavilov tallied up a list of more than six hundred crops that had their greatest number of varieties in montane landscapes and noted that those same regions were rich in indigenous languages and wild biodiversity as well:

> The centers [of diversity] for most cultivated plants turn out to be regions which are the scenes of a vigorous speciation process. Naturally, it was in these regions that early humans flocked, for their floras are rich in edible species. . . . It is therefore very probable that mountainous regions are not only the primary centers of varietal diversity among crops, but also the most ancient nursery grounds of agriculture.

Recently, archaeologist David Harris pointed out that Vavilov should properly be considered the first to recognize most of the important centers of agricultural as well as wild biodiversity, but noted that Vavilov's assertion that these were the natal grounds of the earliest domesticated plants and agricultural practices has not been validated:

> Ever since Vavilov himself equated centers of crop diversity with the homelands of agriculture, there has been conceptual confusion between the two. Despite the massive investment of archaeological effort since 1950 that has gone into investigations of early agriculture, we cannot be confident that plants were domesticated and agriculture developed earlier in the so-called nuclear centers than in other regions of the world. . . . It is time that we conceptually decouple the world pattern of crop-plant diversity that Vavilov so brilliantly demonstrated [from the notion that those nuclear centers are necessarily the places] of the origin and early development of agriculture.

In this narrative, I will call those "nuclear" regions centers of diversity, for that is how they have been widely understood by geneticists, conservation biologists, and biogeographers. Although Vavilov's mapping of those centers had a profound impact on most Russian scientists during his lifetime, his maps did not reach many English-speaking scientists until after 1950. That was when prominent scholars and field scientists such as Carl Sauer, Jack Harlan, J. G. Hawkes, and C. D. Darlington first included modifications of

Vavilov's maps in their own influential works regarding agricultural origins and dispersals. Those scientists clearly understood the immense utility of Vavilov's work in terms of plant exploration and genetic conservation, for although those centers cover far less than a fifth of the land surface of the planet, they harbor a disproportionately high percentage of all wild and domesticated plant diversity.

Roughly a half century after the initial recognition of those centers of diversity by scientists, many of the same regions received a conceptual "makeover" that attracted renewed attention to them. After British environmental analyst Norman Myers released his own map of areas rich in biodiversity in 1988—based on both plant and animal distributions—those centers began to be known as "hotspots of biodiversity" by Conservation International and "Global 200 ecoregions for biodiversity" by the World Wildlife Fund. A new generation of biologists, geographers, planners, and policy makers began to think of such regions as priority "targets" for conservation. Ironically, few of the contemporary enthusiasts of such regions acknowledge their intellectual debt to Vavilov, for their hotspots and critical ecoregions clearly circumscribe many of the same places that Vavilov first mapped. Curiously, even though Vavilov lacked good geographic data on animals of all kinds, and Myers initially lacked much data on insects, their first approximations of diversity-rich regions have largely stood the test of time.

Nonprofits such as Conservation International, World Wildlife Fund, and The Nature Conservancy have promoted the concept of hotspots to fund their purchase and protection of lands harboring high biodiversity. Unfortunately, as anthropologist Mac Chapin has revealed, there have been cases in which nonprofit programs have attempted to either purchase the lands from their original stewards or comanage them for bio-prospecting. In a few well-documented cases, these changes in land management have inadvertently been at the expense of the indigenous farmers and forest cultivators who have long managed the diversity. A recent critique of this approach in *Science* magazine argued that "the bottom line is that biodiver-

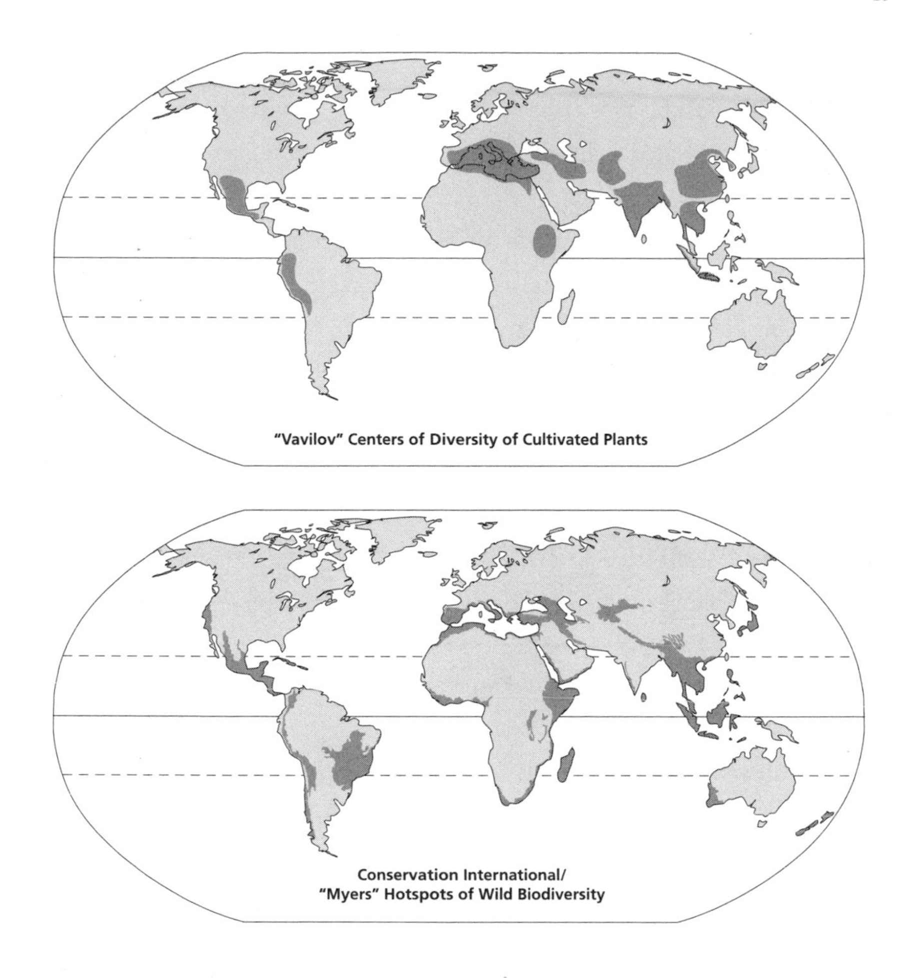

sity will only be conserved if local people and interests want to save it for both ethical and broadly utilitarian reasons."

As if foreshadowing the current debate, Vavilov began to articulate a groundbreaking principle in plant geography that implicitly *includes* human cultures rather than ignoring them, a principle that he explained in *Phytogeographic Basis of Plant Breeding*, published in 1935:

> *The distribution of plant species on earth is not uniform.* There are a number of regions in the world which possess exceptionally large numbers of varieties. . . .

> As far as the crops [concentrated in each of those exceptional regions] are concerned, it is possible to witness there the great role played by Man in the selection of the cultivated forms best suited to each area. [Emphasis added.]

Vavilov not only paved the way for biogeographers to map the patterns of biological diversity, he was also the first to note that biodiverse regions harbored considerable cultural diversity. The individuals participating in those diverse cultures expressed themselves through many indigenous languages and dialects that encoded an enormous wealth of traditional ecological knowledge. Vavilov and Alphonse de Candolle were the first two biogeographers to use linguistic data from diverse cultures as aids in discerning where certain crops originated. In addition, Vavilov proposed that members of various dialect groups sometimes selected their crop varieties for different purposes and environments, and named them differently to encode those distinctions.

Those and other insights first surfaced not in Vavilov's scholarly writing accomplished back in Saint Petersburg, but in the volumes of notes generated while he was visiting fields and gardens in the course of his travels to five continents. Those field notes often provide inventories of the seeds he collected from particular markets or fields; they often give the precise elevational range over which a crop species or variety occurred in a particular valley that he visited. Ultimately, Vavilov's field observations may be seen to be as valuable as the seeds themselves, for they record the historic conditions in which the seeds were grown—conditions that in most cases have dramatically changed over the last nine decades.

Several scientists, historians, and conservationists have written about Vavilov, his rise and his demise, but none to my knowledge have used his field notes to guide them through the same landscapes to ascertain just what kinds of changes have occurred there. Vavilov's photographs and his journals provide extraordinary "snapshots" of the agricultural diversity extant in a particular place and time. They provide a benchmark by which we can measure the rapidity and severity of ecological, agricultural, and socioeconomic changes in particular rural landscapes over many decades.

By retracing Vavilov's steps, and rephotographing the fields, plantations, and markets he visited, it is possible to determine whether (or how much) genetic change has occurred. By interviewing local farmers, ecologists, climatologists, and historians, it is also possible in some instances to discern *why* changes have occurred. In short, Vavilov's notebooks and field photos can remind us not only of where our food comes from, but also of how and why fundamentally important resources are vanishing. Such knowledge may also help motivate us to do whatever is within our power to curb such losses before it is too late.

When one walks the streets of Saint Petersburg, hearing the stories of the dreadful siege it suffered more than a half century ago, it is easy to imagine a world where hunger rages while the great historic legacies of humankind become increasingly imperiled, caught in the crossfire between political powers. I offer this journey in Vavilov's footsteps in the hope that we can more fully value such legacies, that we can more effectively reduce the human suffering associated with such hunger, and that we can sow the seeds of a more permanent peace. But like the seed keepers of Leningrad in 1941, we are in a race against time.

CHAPTER TWO

The Hunger Artist and the Horn of Plenty

Nothing so signifies the bounty of nature as the horn of plenty, the cornucopia. The enduring image of a goat's horn overflowing with fruit, flowers, and grain has been used since the time of the ancient Greeks and up through Vavilov's era and our own to signify abundance, prosperity, and food security. It is said that when the nanny goat Amalthea suckled Zeus, he accidentally broke off one of her horns, which thereafter became filled with nourishment of all kinds. As farmers have seen it, the cornucopia is the agricultural bounty that Vavilov passionately pursued across five continents.

More than any other person before or since, Nikolay Vavilov sampled that bounty in one landscape after another, from interior China across the Silk Road to the Pamir highlands and the Afghan plains, eastward through the semiarid steppes of Iran, Iraq, and Georgia; southward through Lebanon and Syria; across the Mediterranean to Egypt, Tunisia, and Morocco; over the Straits of Gibraltar to Spain, Italy, and Greece; then down into Africa to the heart of the Ethiopian highlands. In the Americas, he surveyed, sampled, and tasted the foods of the Amazon and the Andean highlands, as well as the potatoes of Chile; he sank down into the lowland tropics of Brazil and Colombia, crossed the Isthmus of Panama to evaluate the native crops of Central America, Mexico, and the Caribbean, before returning northward to the United States, where he traveled from California and Arizona to New York and Virginia. Whenever I have had even a fleeting chance to see the colorful variety of fruits and vegetables spilling onto a table from a horn of plenty from any of these countries—the evergreen blues of kales, the glossy greens of mints, the russets of apples, or the purple hues of potatoes—I have been left with a visceral sense that such variety is intrinsically good for us, not just a feast for our eyes. It is a pleasure to our senses, filling us with flavors and fragrances, and essential to our nutritional health as well.

The links between food diversity, bounty, health, and *food security*—the capacity of a community or nation to stave off malnutrition and famine—were first fully articulated in the seminal work of Nikolay Vavilov. Even though the Irish potato famine should have awakened the world to the fact that a narrow genetic base for staple crops is unlikely to carry all the forms of resistance needed to fend off pestilence and plague, Vavilov had to articulate that in principle, or the potato famine would still be considered a special case, an exception. To this day, the notion that the biological diversity in our food-producing landscapes serves as a bet-hedging strategy that contributes to our long-term food security is lost on many people, even some who are deeply concerned with issues of hunger and malnutrition.

And yet it is not lost on farmers, who have seen plagues devastate some mono-cropped fields while barely damaging fields of mixed crops. The corn blight that first hit China and then the United States in the 1970s still did not convince enough of the policy makers and political pundits that genetic uniformity in our major crops can lead to famine or political instability, even though those episodes scared the wits out of most crop geneticists and plant pathologists. Perhaps the large bureaucracies of China and the United States made it look like they had the problem under control, but that was certainly not the case in many smaller African, Latin American, or Caribbean countries. And yet what has saved many Africans, for example, from drought and famine time and time again is, as Andrew Mushita and Carol Thompson have documented, "their own production of [traditional] foods. They are saved by the biodiversity of their food sources; many of the 2000 indigenous food crops are still preserved in rural areas [and their] traditional ecological knowledge designates some highly drought-resistant plants to be eaten only in times of dire need. Botswana alone has 250 plants that are used specifically as 'famine food.'"

Such a high estimation of the importance of biodiversity to food security is not merely the opinion of a couple of African agricultural policy experts; it was also affirmed by the FAO on World Food Day in 2004:

> [The loss of biodiversity in food systems] also affects communities in the poorest regions of the world whose well-being relies on the diversity of nature—plants, animals, forests, water. Poor people are extremely dependent on these diverse resources, and are increasingly vulnerable to the loss of biodiversity. However, poverty often forces people to give priority to immediate needs, and use resources in an unsustainable way.

Whether due to the actions of the poor or the rich, or both, diversity available in the world's food systems has been diminished in the ninety years since Vavilov's first assessment of that agricultural bounty. The causes of that genetic erosion or biodiversity loss are many and include the wholesale replacement of many traditional food varieties by a single cash crop; the conversion or fragmentation of agricultural landscapes by industrialization and urbanization; the usurpment of waters formerly used for crop production for other uses; the loss of traditional seed saving knowledge among the rural populace as farmers become lured by advertising into buying hybrid seeds; and the banning of local production of traditional varieties by plant patenting legislation and free trade agreements. The FAO estimates that about three-quarters of the genetic diversity of agricultural crops has been lost over the last century, and that out of 6,300 animal breeds, 1,350 are endangered or are already extinct. Both livestock and seed diversity have been diminished.

Oddly, many environmentalists who champion the conservation of imperiled habitats in the global hotspots of biodiversity often overlook this loss of food varieties and appear not to realize just how biodiversity buffers people from unpredictable weather, disease, and pestilence. They often adhere to a hands-off attitude regarding nature's diversity that favors strictly protected areas such as national parks over the more interactive conservation and restoration strategies commonly used in most community-based reserves. Perhaps those preservationists forget that their sustenance comes from the intimate ways in which much of the world's diversity has been held for centuries in the gentle but work-worn hands of farmers. Even the more erudite among them may not remember that the very concept of "centers of diversity"—on which many conservation strategies are based—emerged not from studies of wildlands, but from studies of indigenous agriculture. It was Vavilov, a Russian agricultural scientist of peasant ancestry—not an armchair botanist or wilderness advocate—who first circumscribed the most biologically diverse regions of the world and became their champion.

Nikolay Vavilov began his career searching through fields filled with experimental grow-outs of traditional varieties of food plants, looking for those that were resistant to plant diseases. He believed that seeding agricultural areas more widely with resistant strains might help to prevent the devastating crop failures that had triggered numerous famines among the peoples of Russia and the rest of Eastern Europe. His initial experiences in screening crop diversity for sources of disease resistance opened his mind to explore the broader world of biodiversity and led to many more realizations, many extended journeys, and many discoveries in the field that benefit humankind to this day.

To fathom how a Russian plant pathologist of the past century changed the way we think about where our food comes from, it may help to know a bit about where the man himself came from. Understanding the milieu out of which Vavilov emerged may help us understand how he was able to grasp so firmly how food diversity contributes to food security, a connection that had slipped by so many others. His contemporaries have referred to Vavilov's disarming charisma, his elegance, his intellectual and linguistic prowess, and his unflagging enthusiasm, but more than anything else they remember his wanderlust. His restless love for learning the patterns of plant distribution enabled him to recognize the geographic hotspots through which we, too, will soon travel, retracing his steps. His early life history appears to have inadvertently prepared him for a professional life as a peripatetic traveler between cultures, countries, and disciplines.

Nikolay Ivanovich Vavilov was born in Moscow on November 25, 1887, during the era in which czarist Russia was at the pinnacle of its conspicuous consumption, its extravagant indulgences, and its miseries. In the final years of the czars, while peasants were expending as much as a quarter of their income just to provide staple cereals and legumes for their families' meager diets, the czar and his diplomatic corps were feasting on delicacies from every corner of the world. As Mark Twain wrote in *Innocents Abroad*, the czar was "a man who could open his lips, and ships would fly through the waves, locomotives would speed over the plains, couriers would hurry from village to village, a hundred telegraphs would flash the word to the four corners of an empire that stretched its vast proportions over a seventh of the habitable globe, and countless multitudes of men would spring to do his bidding. . . ."

When the czar arranged a feast for visiting diplomats, the gluttony would continue for hours, if not for days. One ambassador to Russia learned that during his sixteen-week stay in Moscow, the royal kitchen generously provided his party with meat from 48 bulls, 336 rams, 1,680 chickens, 10 deer, and 112 geese, plus 11,200 eggs and 350 pounds of butter! In exasperation with how such inequities in food availability were starving hundreds of thousands of people in the countryside, Count Leo Tolstoy and his daughter opened soup kitchens in village after village of Ryazan province. But when Tolstoy looked back at their two-year effort to make food more equitably distributed in Russia, he realized that because the root causes had not been affected, he could only acquire food of inferior quality for the kitchens; in essence, he despaired that they were "distributing the vomit, regurgitated by the rich." Nikolay Vavilov's father was one of the fortunate few who left such misery behind, so that his children could be raised with greater access to both food and educational resources than most Russians could muster. Although he had been christened Ivan Il'ich Il'in not long after he was born, Nikolay's father later changed the family name to the more urbane Vavilov and left his rural identity behind him. For many years before his move to Moscow, however, growing up in a family of muzhiki peasants in the village of Ivankovo, he had faced the daily gnaw of hunger. It was common for families of his village and of villages throughout much of nineteenth-century Russia to have to ration their cereals, potatoes, and vodka for several months prior to the next harvest, to dilute their stews into weak soups, and to glean wild greens and mushrooms from the roadsides in order to have salads or stuffings for their breads.

Ivankovo, situated some 130 kilometers north of Moscow on an old trade route that delivered farm products into the capital, was a place of biting cold in the winter and blistering heat in the summer. Vavilov's grandfather and great-grandfather were serfs who had had their own bouts with famine. Indeed, few Russian families other than the Romanovs in the czar's palace were immune to hunger during that or earlier eras. Vavilov was born into a milieu shaped by periodic, catastrophic famine.

Between AD 1500 and 1700, at least 150 famines ravaged Russia and other regions of Eastern Europe. Russia alone suffered more than 100 hunger years and more than 120 full-tilt famine years over a thousand-year period beginning around AD 873. One of the worst of those famines, in 1873, had an enormous effect on Tolstoy; it shaped his literary ponderings, his choice of a vegetarian lifestyle, and his political actions for the rest of his life. Although Vavilov's work fighting famine may not be recognized today by as many who recognize the works of Tolstoy, it ultimately had larger repercussions on Russia's destiny.

Living through the famine of 1891 and 1892 had a profound effect on Vavilov just as the 1873 famine had on Tolstoy. It was not simply a historical fact that Nikolay learned about as he was growing up; it tangibly determined what was or was not available on the kitchen table of his home during his boyhood in Moscow.

Nikolay was only four years of age when that famine spread from the Volga to devastate rural villages throughout central Russia, threatening the health and survival of fourteen to twenty million people around him. Four hundred thousand Russians had died, and many of the rest were saved by the bare-bones provisions left in their larders. By the time he was just six years old, Nikolay had witnessed starvation and had seen relatives debilitated by malnutrition.

The famine had begun in a most unremarkable manner, with a dry fall in 1890 forcing a delay in the sowing of winter wheat and barley. Such delayed plantings were not unheard of, for they were actually one of many strategies that farmers used to minimize their losses in the face of unusual weather. By sowing their seeds later than usual, they hoped that their crops would be ensured higher survival rates.

Their hopes waned when severe cold arrived much earlier than anticipated in the winter of 1890–91. Then came snows too light to provide a protective blanket to buffer the sprouts from below-freezing temperatures. The young plants of most of the belatedly sown cereals froze to death, collapsing onto the icy ground. Flocks of birds picked at the flattened stalks of the cereals while the peasants looked on, wondering where their grain would come from if the next harvest passed without filling their stores.

By the arrival of spring, most individual farmers realized that they had lost most if not all of their grain crops. It was not until several months later—in June of 1891, when each farmer should have been selling tons of cereal to grain merchants—that it dawned on government officials that they would soon be confronted with an enormous food crisis. A nearly complete crop failure had occurred across some 900,000 square miles, centered on the Volga River Basin but reaching out in all directions into the Russian hinterlands.

Tragically, instead of distributing the yield locally, what little grain there was to harvest was immediately purchased by merchants for export to regions of Eastern Europe that had also been afflicted but had more purchasing power. By late November of 1891, most of the cereal reserves in rural granaries and urban storehouses and mills had been depleted. As bread became a scarce commodity even in Moscow and Saint Petersburg, many Russians prayed for a larger harvest the coming summer, to stave off disaster.

Their prayers were not met. The seed that they had held back had finally been sown, but its seedlings formed only a modest stand a few inches tall by early winter, and by the spring equinox of 1892, cold, drenching rains funneled down from the frozen uplands to scour the seedlings out of the fields in the bottomlands. Ice cold waters surged over the riverbanks and drowned field after field. When the fields were drained off in late spring, there was not a stand of wheat left that was thick enough to make harvesting its grain worthwhile; the farmers just let their livestock out onto it, to consume it as fodder.

The people on the land soon found that they, too, would have to consume straw and chopped weeds in their breads, blintzes, and porridges in order to stay alive. The summer of 1892 began far too hot and dry to plant the typical spectrum of vegetable crops; what did get put in the ground withered and died when five straight months of rainless weather followed. Not only were the grain reserves exhausted, but now all the warm-season vegetables—from asparagus to zucchini—failed to yield enough for the fall harvest.

The proletariat—especially the laborers in factories—became agitated by rumors that the czar had allowed some wealthy Russian merchants to sell off the remaining stores of wheat and oats, barley and buckwheat left in the country. Workers refused to eat any more of the "hunger bread"—from which they had picked pieces of tree bark and cockleburs—placed before them in the cafeterias and soup kitchens. Later recalling why he had become a dissident, Lenin described the typical hunger bread of that era as "a lump of hard black earth covered with a coating of mold." W. C. Edgar, another witness to that famine, wrote that that poor excuse for bread "was so disgusting in smell, taste and appearance that it is difficult to imagine that mankind could be reduced to such extremity as to be forced to eat it."

It was into this era of calamity and inequity that the world's greatest agricultural scientist had been born and the seeds of the Russian Revolution were sown.

Nikolay's father, Ivan Vavilov, had left his family heritage of farming behind to work in the merchant class of Moscow. By the time he married and started a family, he had already moved through menial jobs and become a shop manager and director for a Moscow textile factory. The company was soon to make large profits supplying clothing and blankets to the army at the advent of World War I, and the Vavilov family's assets grew substantially during wartime.

Unlike many others who left the land for the factory during this era, Ivan was able not only to comfortably support his wife, Alexandra Mikhailovna, and children, but also to provide an excellent education for his two daughters and two sons. Among his and Alexandra's heartbreaks was the loss of three other children, all of whom died in their infancy. Such were the times. Some claim that Ivan eventually amassed a fortune but lost much of his wealth during the revolution.

The premature deaths of their siblings no doubt had a deep emotional effect on the surviving Vavilov children. It may partially explain why they sought cures for society's problems through training in the sciences. Though their father encouraged them to follow him into business, all four chose instead to become scientists of sorts: Alexandra, a physician; Lidia, a bacteriologist; Sergei, a physicist; and Nikolay, a plant pathologist and geneticist. Russian science journalist Mark Popovsky elaborated on why the Vavilov children might have chosen the white lab coat over the merchant's business suit:

> The most persistent idea of that period was a faith in a scientific evolution of society. The twentieth century began with a series of major scientific discoveries, and the more enlightened of Russian society became firmly convinced that science was going to bring the world great blessings. From the very beginning

> Russians conceived of science as a branch of social service, a field of activity primarily to benefit the peasant and all those who labored for scanty rewards. . . . Biology, whether it was connected with agronomy or with medicine, was regarded as a department of science concerned with the feeding of the population, with bread, and with people's health.

At the dawn of the twentieth century, scientists were on the verge of understanding how deeply heritable factors affected the health of humans, their livestock, and their crops. During that historic moment, there may have been an overriding political reason for Russian scientists to become intent on discerning what might be inherited susceptibility versus environmental vulnerability to disease: The very survival of the Russian Empire seemed to hinge on whether the only male heir to the throne could survive a genetic malady that had apparently run rampant through his mother's family.

Crown Prince Alexei, the only son of Czar Nicholas II and Empress Alexandra, had been born in 1904 with a debilitating blood disorder that afflicted him almost immediately after birth. He and all other males in the Romanov family—descendants of Queen Victoria of England—suffered from internal hemorrhaging, bleeding and pain at the joints, abdominal swelling, and a pallor symptomatic of a kind of anemia. The symptoms appeared to be an inherited disorder that robbed Alexei and his male kin of their ability to fight disease or injury.

Although the knowledge of Alexei's disorder was suppressed immediately after his birth, a book was released across Europe and Russia in 1914 that linked political destinies to genetic maladies. Princess Catherine Radziwill's *Behind the Veil at the Russian Court* suggested that an inherited blood disorder might leave the Romanovs without an heir to become the next czar. Many of the genetics texts over the following few decades assumed that Alexei suffered from a hemophilia passed on from his mother, but it is now believed that he inherited a sex-linked deficiency, located on the G6PD (favism) gene. This genetic disorder made him vulnerable to anemia, and to the excessive bleeding that Rasputin helped him through when he was a boy. If he did indeed survive the massacre of his family, as some have suggested, this disorder would have inevitably led to the chronic leukemia that he reportedly suffered as an adult in exile.

Recognition of sex-linked genetic deficiencies was something altogether new for the public to consider, and it became clear that if the newfangled science of Vavilov's era was correct, the disorder of the Romanov males could disrupt the passing on of the empire to the next generation. The crisis in power in Russia was, of course, not averted by this scientific discovery. Instead, the second revolution among the Russian people, in 1917, took the Romanovs out of control and brought the Communists to rule, just as the Vavilov siblings began their careers. When Alexei's first bleeding episodes were made public, Nikolay was barely nineteen, but that calamity may have brought him toward a career that explored the relationships between genetics and disease.

Perhaps the career chose him. He had waffled for a while, unsure whether he wanted to be a physician or a plant pathologist. While still a student, he complained in a letter to his family that he was adrift "without a rudder . . . [and with] no definite, well-defined objective . . . [any] specific goal is still veiled in the mist." Just before he was close to settling upon medical school "by default," however, news hit that Russia was experiencing the worst grain yields since the time he was born. As the first, and largely unsuccessful, revolution against the czar flared up in 1905, the Russian Empire was again on the brink of famine. Wheat yields plummeted, as did yields of rye and oats in the Russian breadbasket, the Caucasian region just east of the Black Sea. At that time, the Caucasus countryside was already filled with social unrest, disrupting the distribution of even the little cereal that had been harvested. In light of these developments (in both grain yields and social unrest), Nikolay Vavilov chose to enroll in the Petrovsky Agricultural Institute, bypassing the medical schools his sisters eventually entered.

At Petrovsky, he spent five years studying with some of the pioneering practitioners of the new science of evolutionary genetics, hoping that it would lead to some fundamental discoveries about how organisms develop immunity to diseases. Perhaps the theoretical insights he gained into plant diseases would eventually aid both medical and agricultural scientists concerned with alleviating human suffering—such were the sort of hopes and aspirations his generation entertained.

It was an exciting as well as a challenging time to be an agricultural scientist. Although Russia was the world's largest producer and exporter of wheat during the early twentieth century, its per-hectare yields were less than two-thirds of those in the United States, less than half of those in France, barely a third of those in Germany, and hardly a fourth of what the Dutch were harvesting in the Netherlands. What's more, wheat diseases and pests regularly halved what the yields per hectare might otherwise have been in Russia. Before Vavilov left the agricultural institute in 1911, Russia's grain farmers were hit with yet another decline in cereal yields that was even worse than the one five years earlier. A pressing challenge for his generation, clearly, would be to do something to alleviate the recurrent famines. Reducing crop plagues and pestilence seemed a sure-fire way to ease the social tensions of Russian society. "I am, above all, a plant pathologist," Vavilov wrote early in his career and said he asked with each discovery, "'How can it be put to practical use? How can it be utilized now, right away, to benefit my country and all toilers of the globe?'"

In the first decades of the twentieth century, agricultural scientists felt that they finally had the technical capacity to prevent crop diseases and plagues from triggering famine and the resulting human deaths, displacement, and social unrest. Although few agricultural scientists today live in fear that a crop epidemic or pestilence may bring our societies to the brink of starvation, the losses caused by those factors are no less troubling today than when Nikolay chose his career path. As one of his earliest Russian biographers, Genady Golubev, reminded us in the late 1970s, in the wake of the corn blight that devastated maize yields in the United States, "Even nowadays, according to the United Nations, plant diseases deprive us of one-third of all the crops grown with so much effort. A whole one third, that is, a billion tons. And this is happening while every third human being on earth is still suffering from starvation. Sixty years ago, however, when Nikolay Vavilov was beginning his struggle against plant pests and diseases, they robbed the farmer of at least half of his crop."

Just as Nikolay undertook his studies of plant pathology at Petrovsky, his professors received a rather astonishing report from England. A British scientist named R. H. Biffen had achieved a major scientific breakthrough in the human struggle against plant disease. His breakthrough drew upon the recently rediscovered principles of inheritance laid down by Mendel, which allowed crop breeders to rather reliably predict the expression of what they called dominant and recessive genes in subsequent generations of plants when an intentional "crossing" of two varieties was made. Biffen used his quantitative approach to genetics to breed resistance to yellow rust into a second generation of hybrid wheat and in that way obtained immunity to the disease in a new wheat variety. The accomplishment was an important first in the history of crop improvement, the agronomic equivalent of reaching the moon. In attempting to place Biffen's momentous achievement in a historic context, his colleague F. T. Brooks made the following argument for integrating plant pathology with both ecology and genetics and carved out a broader path for Vavilov to explore:

> It may be argued that the future control of plant diseases lies in the hands of the plant breeder. This is only partly true [because] all Nature being in a state of flux . . . it must be remembered that pathogenic organisms themselves change, and, with increasing virulence, may attack varieties of crop plants hitherto resistant. It is particularly the province of the plant pathologist to ascertain the conditions of growth of cultivated plants which are least favorable to attack by parasites, and to prevent disease by applying the methods of plant sanitation. The best results in the control of plant diseases are likely to be achieved by the mutual co-operation of plant breeder and plant pathologist. . . .

Brooks thus made it clear that both genetic and ecological factors must be taken into account if ever-evolving plant diseases were to be controlled. In essence, Brooks, Vavilov, and their peers became the first generation of scientists to place inherited immunity and disease susceptibility in the context of both evolutionary genetics and environmental influences on gene expression. To understand why a crop was susceptible or immune to a particular disease, it was critical to understand its current ecological setting, as well as those settings in which its ancestors encountered various stresses. Vavilov thus realized that knowledge of the origin and diffusion of a crop was not mere historic trivia; it would help him search for places where resistance to both diseases and pests had developed. This evolutionary perspective gave Vavilov the intellectual license to seek out crops and their wild relatives in the areas of the world where they emerged and then diversified.

Because both genetics and field ecology were disciplines in their infancy during that time, Vavilov was fortunate enough to be able to absorb their advances in knowledge as if they were merely two sides of the same coin. In the simplest terms, he was able to grasp how inheritance interacted with the environment to shape the destiny of the crops he was studying. Through their courses in both physiology and bacteriology, students of his time were encouraged to observe the effects of genetic as well as environmental factors on the whole plant. Vavilov's mentors encouraged him to integrate the latest theories of natural and artificial selection of plants, but at the same time, they nurtured his capacity to make his own detailed observations, not simply in a laboratory but also in the field. In that sense, he had an advantage over many of today's lab scientists who have never witnessed a crop in the environments where it evolved; Vavilov was able to observe how stresses affected a crop where it had been grown traditionally for centuries and see how it adapted to such stresses.

For Vavilov the student, the "field" meant two things: (1) the experimental plots that his Petrovsky professors maintained on the outskirts of Moscow, where they selected for plants with superior resistance to disease; and (2) the traditional fields of peasants out in the larger landscape, where any advances scientists made had to be tested and found worthy of adoption by farmers. For a Russian to be a field scientist in revolutionary times meant that he was responsive to the proletariat's needs and worked in solidarity with the peasantry, the salt of the earth. But doing science in the field also had a hint of adventure imbedded in it, for it allowed one to roam freely across the earth, looking for seeds, "cross-pollinating" ideas with others from different backgrounds, and, through such intellectual hybridization, gaining fresh insights of potential benefit to humankind.

Vavilov on early field expedition in Turkmenia.

A focus on fieldwork was also an assertion—especially to his brother Sergei, the physicist, and his sisters, the microbiologists—that the real work of agricultural science took place not in a fancy lab, but in the soil itself. And so, before his formal undergraduate education was over, Nikolay was laboring long hours planting, weeding, and taking notes among the experimental plantings of his mentors. He typically awakened as early as any of the farm laborers and assisted them in their toil, but when they stopped for a rest, he worked on, searching for indicators of disease—or immunity to it—among the crops they were harvesting. His hands and his head seemed to work in synchrony, so that whatever his fingers held, his mind registered in the larger scheme of things.

As a social revolution of unprecedented proportions seemed more and more imminent between 1915 and 1917, Nikolay struggled to show his peers that he was not "above" the proletariat and peasant classes—he was willing to engage in the same physical work that less-educated comrades would do. Yet, at the same time, he set goals for himself that he believed would someday make all of their lives easier.

When he was not planting, weeding, cutting, winnowing, or weighing harvests from the experimental fields, Vavilov took flight to a larger field. He broke away from Moscow to begin what would become a lifelong obsession: taking long excursions into the countryside to observe how traditional agriculture functioned in the fields of peasant farmers. While he was still at Petrovsky, he initiated his own studies of ancient agricultural practices in the culturally diverse Caucasian region just west of the Black Sea, which divides Europe from Asia. There, among the muzhiki—men and women cut from the same cloth as his forefathers—he witnessed farming practices that had endured for many centuries and nurtured hundreds of locally adapted crop varieties. With their baggy pants, their thick wool sweaters, and their colorful hats, these peasant farmers met him in their orchards and hay barns with sickles in hand and kindly responded to the myriad questions that sometimes spilled out of this intense young man. Nikolay was attempting to see where among their fields plant diseases first cropped up, and how the muzhiki dealt with them through their traditional ecological knowledge. He could see with his own eyes that the damage inflicted by certain diseases was greatly reduced when those diseases encountered a polyculture, or mix of locally adapted varieties, rather than a monoculture. Rather than dismissing farmers' local knowledge as "unscientific," he found there was much to be learned from it.

Perhaps his respect for peasants had been fostered in the field excursions taken by his classes at the gymnasium, where he had been given some training in the kind of regional ethnographic documentation that Russian explorers had practiced in exotic lands. He had particularly loved the ethnographic excursions and attended lectures by great explorers, anthropologists, and geographers at the Moscow Polytechnic Museum. Now he found himself documenting the agricultural expertise and place-based practices of people much like his own grandfathers.

Vavilov also adopted the habit of collecting specimens of seeds and flowering stalks from the crops of the farmers, just as his teachers had taught him to do with the wild endemic plants that they encountered on their excursions into the mountains. As a child, Nikolay had shared a herbarium in his home with his brother and two sisters. Collecting plants for it was one of the activities he relished; he botanized for pleasure not only with his siblings, but also, later, with his wife when they were together. That naturalist's habit helped to ground him throughout life; perhaps that's why he felt so much sorrow when his Western European plant collections en route to the herbarium went down with the ship they were on in World War I. At Petrovsky, however, he learned to go beyond the mere collection of dead and dried plant specimens for mounting; he ventured into saving seeds.

Because they were living, respiring, reproductively viable organisms that could be regenerated for posterity, seeds became the subjects of Vavilov's desire and the objects of his scientific inquiry. He knew, of course, that they were more than mere "germ plasm"—the regenerative tissue of higher plants. They were also a source of food for abating famine, and they were food for thought. In Vavilov's time, scientists could not see genes; they could only infer their existence by looking at the diverse seeds that embodied their presence. Vavilov found that he could measure the frequency of certain characteristics in living seed lots to ascertain their expression of dominant or recessive genetic traits. (Later, he would grow out samples from the seed lots to confirm the inheritance of their physiological tolerances and disease resistance.)

Before he had reached twenty years of age, Vavilov began what became the largest collection of seeds that the world has ever known. Many of his seed lots were accompanied by field collections of the vegetative and reproductive structures of the plants, which he kept in a herbarium, most of which survive to this day in Saint Petersburg. Like Thoreau, Vavilov had faith in a seed.

Yet Vavilov collected, compared, and conserved far more than the plants themselves; he was just as intent on recording the native names, uses, and lore found among the various agrarian communities he visited. That information provided each fruit or seed he collected with an ecological, cultural, or culinary context. He would never have been able to pick up so many clues about a plant's immunity to disease or resistance to post-harvest insects if he had spoken only Russian. As if preparing for a life of wanderlust, Vavilov had somehow become the quintessential polyglot at a rather early age. Before leaving the gymnasium for Petrovsky, he had already developed his reading skills in German, English, Latin, French, and Italian. While at the agricultural institute, he took advance tutorials in English, so that by the time he first visited London in 1913, he had the capacity to converse with the finest scholars who taught there. By the time of his death, he was so conversant in some fifteen languages—including Farsi, Turkic, and Amharic—that he fired any translators in the field who could not give him accurate running summaries of his conversations with the scientists and farmers hosting him in various countries.

At night, after the conversations had terminated, Vavilov stayed up late recording and comparing various indigenous names for the folk varieties of crops he had seen that day. He discerned minute but significant differences between native names for the same crop and noted which loan words gave insight into the cross-cultural diffusion of particular crop varieties. Although a famous agricultural botanist father-son team—the de Candolles—had casually used loan words to trace the diffusion of certain crop species from one culture to another, Vavilov took the exercise to a deeper level. As his colleague N. A. Maisurian noted about Vavilov's early treatise on the origins of rye as a weed among wheat and barley, "This work had the form of a beautiful *etude*, describing the 'original' history of a famous and widely-cultivated plant. He first showed the possibility of applying linguistic analysis to botanical research. After Vavilov . . . this method was used by [many] other scientists."

From his notes, you can see how Vavilov intuitively sensed that if farmers in a community gave two closely related varieties different names, they almost always isolated those varieties into two distinctive microenvironments, allowing natural and cultural selection to shape their subsequent evolution in different ways. When he traveled up an agricultural valley whose inhabitants had their own dialect—as he did in several valleys in the Pamirs of Tajikistan—he observed what farmers called their lowland versus highland varieties, and he assessed the performances of both varieties along the steep altitudinal gradients he encountered there.

In essence, Vavilov's extraordinary linguistic skills and geographic explorations led him to anticipate that there might be relationships between the linguistic diversity and the agrobiodiversity of a region that would merit more protracted attention. Did the different styles of linguistic marking in two adjacent cultures lead farmers from each culture to select traits in similar crop varieties in different manners? Did the mere act of naming a crop variety ultimately lead to its divergence from close relatives? Vavilov was raising such questions fifty to seventy years before they became of intense interest to linguistic anthropologists and ethnobiologists.

To be conversant in genetics, ecology, evolution, linguistics, and geography at such an early age, Vavilov not only had to be bright and well mentored, he must also have been driven by an unquenchable thirst for knowledge. His friend Carlos Offerman, who stayed in the Vavilov household over many years, put it simply: "He had extraordinary vitality, was always on the go from early morning until late at night; three and a half hours daily was all the sleep he needed. . . . The only time I have known Vavilov to be sick . . . he was staying home that day with the flu. Even on such an occasion, he was surrounded by books, using the time to work. He received us both jovially, and discussed at length the problems of the Argentine." At Petrovsky Vavilov dove with intensity into several years of study of plant immunity to both diseases and pests, even while the Russian Revolution of 1905 raged in the Presnya District all around him. In 1908 he completed his bachelor's thesis, in which he addressed the ecology of field slugs as pests in winter crops around Moscow. The next year he published his first scientific essay, "On Darwinism and Experimental Morphology." He took two more winters to accomplish additional experiments on crop resistance to slugs in the fields and gardens surrounding the city, and then he entered his study in a science fair competition at the Moscow Polytechnic Museum. In essence, Vavilov was beginning to muse upon ways that the very shape and form of a crop plant repel (or invite) certain pests and insects; that kind of attention to detail greatly aided him later on, as he visited tens of thousands of fields on five continents, selecting seed from plants that he felt might have resistance.

Vavilov's study of field slugs was awarded first prize in 1911. The next year he married his first wife, Ekaterina, also a student. Soon after, their mentor, Dionazas Leopo'dovich Rudzinkas, encouraged them to move on, to study abroad, for Nikolay in particular had already outgrown what Petrovsky professors could offer him. "I am ashamed when you call me your teacher," Rudzinkas told the young Vavilov, "[for] I borrowed from you many more times than you borrowed from me."

By 1913, Nikolay and his equally talented bride, Ekaterina, had accepted their mentor's admonitions to move into the larger world, where their talents could be further nurtured. They sailed to England, France, and Germany, where for the next two years they were tutored by some of the finest scientists and seeds men of the twentieth century. In England, they both spent time at the John Innes Horticultural Institute at Merton College at Cambridge University and at the University of Reading. At the horticultural institute, they began a decade-long friendship with zoologist William Bateson, a pioneer in the modern science of evolutionary genetics. Bateson first coined the term *genetics* in a 1905 open letter to other scientists and played a key role in fostering the use of concepts such as the gene pool. That a laboratory zoologist was the dean of horticultural science at Merton both amused and inspired Vavilov:

> The title "horticultural" was irrelevant to reality at the John Innes Institute; it was actually the Mecca and Medina to geneticists of the entire world. . . . One could only marvel at the diversity of subjects approached by the geneticists. . . . Wheat, flax, rabbits, hens, canaries, straight wings, begonias, tobacco, potatoes, lion's mouth, plums, apples, strawberries and peacocks, all were research objects

> studied. . . . Physiological questions were studied in addition to genetics. Topics were sometimes chosen without a [preconceived] general plan. Bateson himself had difficulty in selecting a subject for me, and much to his pleasure he allowed the author of these lines to continue at Merton his work on the immunity of cereals.

On a break from their work with Bateson in 1913 and 1914, the Vavilovs traveled to France. There they visited one of the world's oldest and largest seed companies, Vilmorin-Andrieux & Co., which already had two hundred years of success scientifically selecting and introducing new grain and vegetable varieties to farmers in the French countryside and in France's many colonies. They then moved overland to Jena, Germany, where Bateson's fellow pioneer in evolutionary genetics, Ernst Haeckel, hosted them.

In August of 1914, as war was breaking over Europe, the Vavilovs returned home to Moscow, Nikolay carrying with him a seed and plant collection that he had gathered from their many stops along the way. He also tried to send back to Moscow a rather large collection of books he had been reading, many of them addressing the emerging field of the genetic and evolutionary origins of cultivated plants. Even before the war was over, those readings propelled him to undertake his first major field excursion for seed collecting in 1916—to the "roof of the world" in present-day Tajikistan—despite the fact that he and his bride had not had much of a chance to settle down. Just as World War I began to wane but Russia boiled over into its second revolution—this one toppling the czar—Vavilov returned from the field to accept a position as professor at the University of Saratov. As he and Ekaterina hunkered down in Saratov, she gave birth to their son, Oleg Nikolaevich Vavilov, on November 7, 1918.

It was not long after returning to their homeland from Europe—when Ekaterina was still holding a baby within her womb—that Nikolay first turned his attention to the genetic origins of cultivated plants and to the geographic birthing grounds of agriculture. In late 1917, he published "On the Origin of Cultivated Rye," but his interest in the topic was not restricted to the histories of various grains. In fact, he had begun to develop some hypotheses that the most productive places to seek out sources of immunity to diseases in crops might be their geographic birthing grounds, which he had begun to call "their centers of origin."

Perhaps—as he later argued in a 1926 monograph, "The Centers of Origin of Cultivated Plants"—those centers still held ancient, diverse forms of crops that had coevolved with pests and diseases over many millennia. It is clear that from an early age Vavilov was thinking about plant pathology in an evolutionary, geographic context, rather than assuming that plant diseases randomly crop up in some sort of vacuum. Perhaps only a few of the varieties of the crops that had developed some immunity to pestilence and plague had been taken from their centers of origin to be cultivated in distant lands. If so, going back to the original source area might be the best place to find additional genes for resistance. If the crop failures that had persisted in Russia throughout his youth were ever to be quelled, perhaps it was time for Russian scientists to visit the birthing grounds of wheat, barley, oats, rye, potatoes, and corn and bring back from them some relatively unused sources of immunity. Of course, such a proposition depended on mobilizing a tremendous effort to both identify and visit those cradles of agriculture that he called the centers of origin.

Even before Nikolay had settled down to rock his son in his cradle, he let Ekaterina know of his burning desire to take on such a wayfaring task. He repeated the words he had written to her when she was still his fiancée: "I will make no bones about my longing—immodest as it may be—to devote myself to the Road of searching . . . the *Eforschung Weg*."

Nikolay's wanderlust eventually strained his marriage with Ekaterina, who spent the next seven years living with her in-laws while Vavilov wandered the earth, searching for the cradles of agriculture. In 1926, she had their marriage dissolved, and with William Bateson's help, she emigrated to Canada, taking their son, Oleg, with her.

Interestingly, Oleg would develop the same passion for wandering in remote mountainous regions that his father had exhibited from early on. Like his father, he developed a passion for both field science and mountain climbing, and he eventually returned to the Soviet Union to pursue both. But, on February 4, 1946—about the time the Western world learned the fate of his father—Oleg Vavilov died at the age of twenty-seven years while making a treacherous ascent of a snow-capped mountain in the Caucasus.

In a way both men died from the consequences of their insatiable wanderlust. But Nikolay's death occurred many years after he caught the bug of wanderlust and after many groundbreaking forays to the nursery grounds of the world's ancient seeds. We follow him now into the field—into many fields nestled in the mountains of five continents.

CHAPTER THREE

Melting Glaciers and Waves of Grain: The Pamirs

In retrospect, it seems altogether improbable that someone might actually have caught the very moment when Vavilov first set out to discover where our food comes from—the centers of origin for cultivated plants that some scientists now call the hotspots of agricultural biodiversity. Perhaps even more miraculous is that the scene, which took place in May 1916, when Vavilov was twenty-nine years old, was so vividly recorded.

Relatively early in Nikolay Ivanovich Vavilov's career, his dreams soared beyond his neat experimental plots scattered around Moscow and out into the crazy-quilt of traditional fields tended by the peasantry, which still fed most of the world. Like a Don Quixote, Nikolay set his sights high, deciding he would try to collect the entire range of food diversity found on all five continents. His cousin A. I. Ipat'ev gives us this account of the day Nikolay began his first sortie to what he would call the centers of origin of cultivated plants:

> One clear summer day in 1916 an automobile—a great rarity then—pulled up to the house. Nikolay Ivanovich came up to me while I was sitting in the garden and greeted me. He was, as always, radiant and gay; only his appearance was unusual and strange. He was wearing a cream-colored summer suit; across his shoulders was a full pack; and on his head was the strangest thing of all, a white hat with a double-brim, which he called a "hello-goodbye" hat. He got into the auto and drove off.

That sunny day in May—on the eve of the 1917 Russian Revolution—Nikolay told his family that he would be away until August on an expedition to collect a famed strain of Persian wheat, and to further develop a seed bank that would contain samples representing all the grain crops in Iran. However, official documents reveal that he had quite a different assignment from the czarist government: to determine why Russian troops at remote garrisons were getting sick on the wheat flour in their rations.

Soon Nikolay arrived at the first garrison near the Iranian border, where he quickly surmised that the soldiers were being fed flour that included not only the ground grains of wheat, but the ergot-infested seeds of a weedy grass named darnel, as well. Ergot is a fungus that attacks the grass seeds, producing toxic but typically sublethal levels of lysergic acid, the naturally occurring drug that later became famous as LSD. Nikolay quickly completed his official assignment by making a simple suggestion to the garrison commander: The men would probably stop hallucinating soon after he bought them some better-quality flour. The commander appeared grateful, so much so that Nikolay assumed he would be given the logistical support to freely wander through Persian fields and gardens, collecting as many seeds and taking as many field notes as he pleased.

But in the very first days of his first full-fledged expedition, he stumbled into a war zone where another squadron of Russian troops had not been briefed on his mission among them. Glancing at this elegantly dressed, well-educated young man, they took him to be a German spy. The soldiers detained and interrogated him for several days before he was able to get word to friends and officials in Moscow, who expeditiously arranged for his release.

Delayed but intellectually undeterred, Vavilov once again pursued his search for the mythic disease-resistant Persian wheat. Oddly, he never encountered a single seed of it in the field, neither on that first trip nor on any of his other expeditions. Nevertheless, he got so much into the groove of collecting seeds and describing the growing conditions of various crops that he simply could not stop. Once he had established his pace of plant collecting in Persia, he set his sights on Kyrgyzstan and Mongolia, forgetting all about returning to Moscow by summer's end. He journeyed further eastward, sending home notes and seeds whenever he could. Unfortunately, his field reports did not give much pleasure to his anxious family back in Moscow; at one point, he wrote that "suddenly a Kyrgyzian uprising occurred in Semirech'e and consequently the route to Mongolia was closed." The only way to get to his destination in the Pamirs was to cross the Demri-Shaurg glacier.

Vavilov in the Pamirs of Tajikistan.

Only later did he tell his family and friends that his Kirgiz guides had deserted him and he had been attacked by a mob but somehow escaped from them on foot, only to be arrested by local officials. Once the local police released him, he was forced to shift his plans; he journeyed southeastward until he came to one of the many spur trails that constituted the Silk Road, the ancient trade route between Central Asia and the Far East. It was on that spur that in late August 1916 he entered into the high, dry, and lonesome land known as the Pamirs of Gorno-Badakhshan. His arrival in the mountains of Central Asia was in essence his initial encounter with the kind of landscape some now call a hotspot of biodiversity, for this modest-sized region contains some 5,500 species of plants—1,500 of them endemic (unique to the region)—as well as 143 species of mammals and 493 species of birds. No wonder Vavilov referred to it as a center for diversity.

Somewhere before leaving the Kirgiz for the Tajiks to the east, a Russian agent in Bukhara arranged for young Vavilov to be given horses, mules, and the assistance of a remarkable multilingual guide. A man of enormous linguistic talent and girth, Khan Kil'dy Mirza-Bashi was a local pasha who wore colorful embroidered robes from the Silk Road, and taught Vavilov how to study the culinary diversity whenever they sat down together for a local feast. Thus Mirza-Bashi not only served as Nikolay's interpreter and guide, but also as his mentor in the tricks of back country travel, which would stand him in good stead for many years. The pasha organized provisions for the six mounts and two sherpa-like equipment-bearers to accompany them on their journey across snow-covered mountain passes and wind-swept glaciers, but even those companions did not necessarily ensure him of safe passage. As Nikolay later admitted,

> The caravan proceeded slowly along a barely passable path, stopping overnight in small villages. The trail toward Garm was cut almost in two by an enormous, almost perpendicular cliff. Later on we encountered many complicated mountain passes, but, very likely, this was one of the worst. . . . The horses had to be led below it through alpine rivers. The guides, spanning a chasm more than a meter wide, formed a human bridge over which Mirza-Bashi and I had to pass. [For] Mirza-Bashi, [it] turned out to be particularly difficult because of his weight.

If that were not tough enough, the delay caused by sending the horses down the gorge, then crossing the chasm between its walls, kept the party from arriving in the first Pamiri village on the other side before darkness fell:

> After the passage through the gorge, a considerable portion of the path went along the edge of a glacier. We had to camp overnight among the rocks. We had not calculated on a night camp along a glacier. The lack of warm clothes forced us to start moving earlier. Almost freezing to death for two days was not very pleasant, and it was alleviated only by a common lowering of expectations, by indifference to all that happened.

At one point along a trail edging a precipitous ridge, two eagles startled Vavilov's mount, and he nearly plummeted into the ravine below. He later noted that this was a defining moment for him as a field researcher: "Such moments steel one for the rest of one's life; they prepare a scientist for all difficulties, all adversities and all things unanticipated. In this respect, my first expedition proved to be especially instructive."

Vavilov and his cohorts had entered the Pamiri highlands of the remote province now known as Gorno-Badakhshan, a legendary highland that in Tajikistan is nicknamed "the roof of the world." Because of its topographic heterogeneity and the historic mixing of many cultures and crops that trailed in on the Silk Road, it is a region where the diversities of both languages and locally adapted crop varieties are obvious. Although this autonomous province is presently part of the (former Soviet) republic of Tajikistan, it has deep historic ties to Afghanistan provinces immediately to the south of it, where the massive mountain range called the Hindu Kush reaches upwards of 7,700 meters.

Much of the Pamiri highlands—the third-highest montane landscape in the world—lies above five thousand meters, forming cold desert valleys between glacier-covered peaks. Those remote valleys are where highland farmers have long worked seeming miracles in coaxing water to trickle into canals along miles of ridges before it moistens the earth in their fields of mixed grains and legumes. In doing that, they developed strategies of plant management that buffered them from the vagaries of the weather and offered them a certain resilience not seen in more industrialized, homogenized agricultural landscapes. The cropping strategies that offer such resilience are now of interest to scientists worldwide, for they may be useful in providing food security to others living in landscapes characterized by rather unpredictable weather and a likelihood of frequent landslides. The frequency and severity of landslides and mudflows appear to be exacerbated by the global climatic shifts from which Tajiks are currently suffering.

Vavilov was perhaps the first to recognize the Pamir highlands as a natural laboratory for crop evolution and resilience. He returned there for extended stays on two additional expeditions, spending more time there all told than in

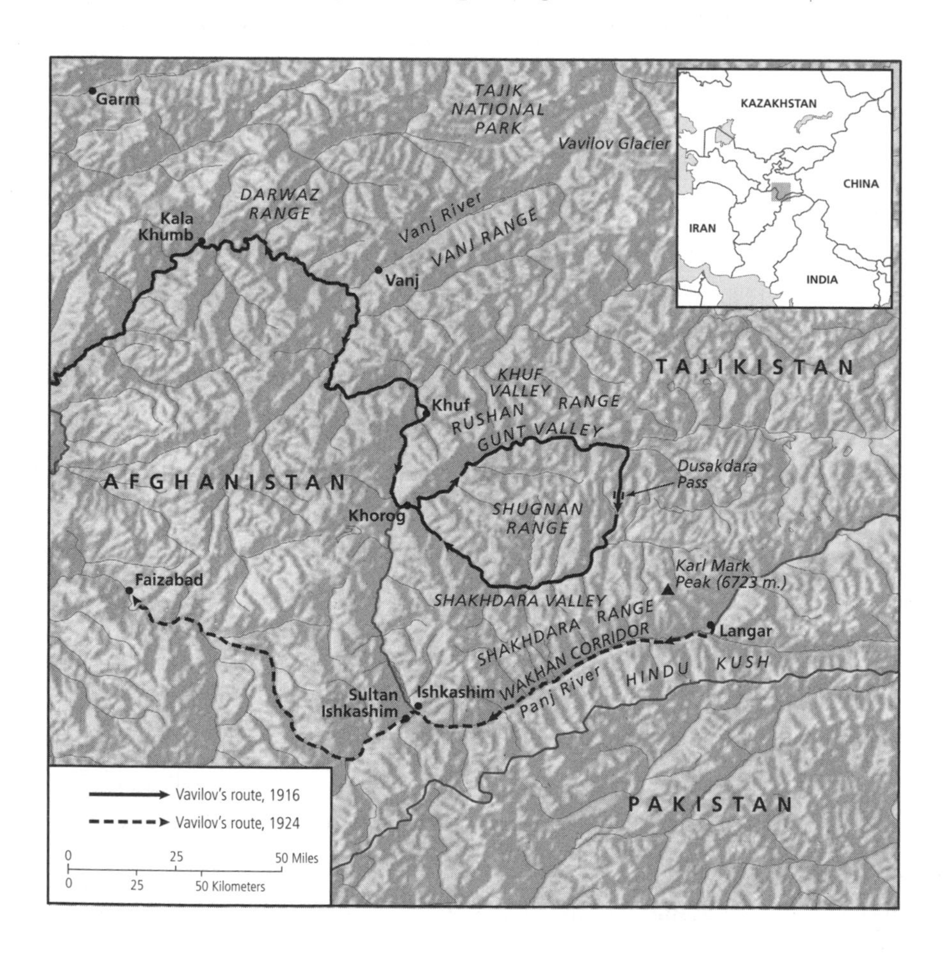

any other region of peasant agriculture he would visit. It was my good fortune to follow Vavilov's path into four of the steepest agricultural valleys in the Western Pamirs, where he had collected seeds some ninety years before. With the seeds Vavilov gathered in those valleys, he founded what became the World Collection of Cultivated Plants that was later housed at VIR in Saint Petersburg. His two-hundred-odd seed collections from the Pamirs included several samples that were later used to successfully breed early-maturing wheats, chickpeas, lentils, and mung beans—cultivars that are feeding people to this day.

It was also here in the Pamirs—while stomping through the muddy rows of verdant fields tucked in below snow-capped cliffs—that Vavilov began to take the meticulous notes that captured my interest even more than the seed collections themselves. In three of the steepest valleys of the Pamirs, for example, his recorded observations are so precise in place and time that we can still use them as benchmarks for assessing changes in climate, crop diversity, farming practices, and food security. In place after place he recorded traditional knowledge of strategies for resilience that might otherwise have been lost and provided snapshots detailed enough to allow a precise assessment of relative loss or persistence of crop diversity through time in the very centers where the crops historically diversified.

When Vavilov first trekked through the Pamirs in August and September of 1916, the glacial melt was slowing its descent off the peaks, and the milky gray rivulets running down the canals were being held in ponds while the peasants harvested the patchwork of grain, legume, and oilseed crops drying on the flattest stretches of arable land. The Pamirs are now known to contain more glaciers than any region south of the Arctic Circle, and one of them bears his name in honor of his first trek there. Those glacier-capped ranges were called the Onion Mountains during the time of Marco Polo, who trekked the onion-scented trails along the River Panj on his way to China. In Vavilov's time, nine kinds of wild onions still grew below the melting glaciers, and many more heirloom varieties of onions were cultivated in the fields, gardens, and orchards where the cool, clear waters of the irrigation ditches flowed.

Over a matter of three weeks near the end of the summer of 1916, Vavilov, the pasha, and their little crew traveled up and down four steep valleys whose rivers drain into the River Panj. The riotous Panj—often raging with snowmelt—continues to this day to serve as the "leaky boundary" between Tajikistan and Afghanistan. Then as now, a bend in the river near the twin villages of Ishkashim is home to contraband dealers, opium smugglers, and other unsavory sorts of border-crossers.

Despite occasional bad company and the logistical difficulties inherent in traveling in such a remote area, Nikolay was at last in his element, collecting seeds of every early-maturing grain and legume that peasant farmers were willing to share with him. Taking barometric pressure measurements that helped determine each crop's elevational range, and making lists of local names for the crop repertoire of each village, he began to discern some geographic patterns in crop diversity on this first excursion that continue to intrigue scientists to this day. He noticed how the ruggedness of the landscape and the frequency of mudslides that blocked roads and trails had, over time, isolated the peoples between adjacent river valleys such that they had gradually developed their own distinctive dialects.

Within each of those dialects—Rushani, Shugni, Ishkashemi, and Wakhi—local farmers had coined their own terms for particular crop species and, more important, for the new varieties that were constantly selected from their own fields. The mere act of naming a newly found variant of onion or apple leads to its isolation and further selection if the novel plant is given special care by an observant farmer. Vavilov somehow surmised that at a larger scale—that is, in the Panj watershed draining the Western Pamirs—linguistic diversity could well have fostered crop diversity.

But the effects of relative isolation were just half of the story. Even farmers in the farthest reaches of the hinterlands periodically venture out to seek other resources. In addition, when those farmers marry women from other valleys, the women often bring seeds of their patrilineal legacy with them to their new home. So once farmers in a particular mountain village had selected and named a new crop variety of some value to them, it was inevitable that it would sooner or later be traded to others in the next valley over, who would adapt it to their own particular agronomic conditions. Call it backyard plant breeding by mass selection, but it was plant breeding, nonetheless, well before Mendel, Bateson, and others fully elucidated the genetic processes through which it occurred. The farmers would mix various locally adapted varieties in ways that generated more heterogeneous plant populations in their fields, then select favorable traits that they wished to keep and share with other farmers. Interestingly, that process has not been arrested in the Pamirs by the arrival of hybrids developed by scientists elsewhere in the world. On the contrary, peasant farmers of various dialect groups on both sides of the River Panj have still found means to exchange their new discoveries and mix them into their broader "portfolio" of seed stocks.

Adoption of those practices has important consequences for food security. The new elements of diversity become harbored in multiple locations, reducing the probability that a single landslide or flood might wipe them out across a region. When disaster struck one valley, the community would send a few of its men out to collect seed stocks from other farmers nearby, particularly from those speaking the various mutually intelligible dialects in the greater Badakhshan region, ranging from southwestern Tajikistan (where I initially arrived) well into northern Afghanistan.

My traveling companions and I decided to follow the seeds southward across the River Panj and the border with Afghanistan to learn more about the traditional means of seed exchange. When we crossed the bridge that traversed the cobble-strewn floodplain of the river, we saw firsthand that the same kind of informal seed trade network had once again begun to function, now that skirmishes had abated along the border between Afghanistan and Tajikistan. After our vehicle rose from the floodplain and passed by terrace after terrace of irrigated orchards and grain and legume crops, we came upon a sprawling marketplace of makeshift stands just outside an old fortress. There we met multilingual traders of culinary herbs and crop seeds who personally transported the products of Pamiri farmers to others at the same elevations as far south as the Khyber Pass hundreds of kilometers away. At first, it was mind-boggling to me that the elderly trader in the turban before me in a small hut on the Tajikistan-Afghanistan border could orally order vegetable seeds or spices from a traveler headed south toward Kabul, who would then relay that order in a dialect of Farsi to others farther to the south, who would then obtain the seeds or spices from a Bengali trader in India. Informal access to such plant diversity—as we were soon to see—offered peasants in the Pamir highlands a kind of resilience in the face of change that no formal government seed introduction program could ever surpass.

Despite all the political and economic upheavals its people have suffered through, the Pamirs can still be considered a land unusually diverse in crop varieties—from onions and garlic to wheat and vetch. That crop diversity, in some ways, now dwarfs the wild plant diversity of the landscape, which is among the highest and driest of deserts made by the rain shadows of towering mountains. Tucked up on a shelf between the Hindu Kush and Tien Shan, the Pamirs receive little precipitation per unit area, but the watersheds are so vast and the glaciers so ancient that they shed enough snowmelt to irrigate the pockets of arable land in the flatter valleys. The snowmelt funneled into their fields allows Pamiri farmers to raise an amazingly rich array of traditional grains, legumes, vegetables, and fruit, as well as forages for fat-tailed sheep, goats, and yaks. Some forty species of crops and six species of livestock enliven the tiny farms of Gorno-Badakhshan.

Yet, even in such a remote and rugged region, traditional farming cultures and their crop assemblages simply do not remain static. Neither does the climate. Even though the montane landscape may superficially look much the same as when both Marco Polo and Nikolay Vavilov first saw it, one feature of the Pamirs is radically different than when their caravans passed on the Silk Road: the alarming rate at which the glaciers plastered up against the Onion Mountains are currently melting. Since Vavilov's passing, some of the glaciers in Tajikistan have lost half of their volume. The largest mass of ice remaining in Central Asia—the Fedchenko Glacier of the Western Pamirs— has lost about two cubic kilometers of ice over just a few decades. Because a quarter of all the water flowing in Tajikistan's streams and irrigation canals comes from those glaciers, their rapid depletion is of great concern to farmers and wilderness trekkers alike.

From the depletion of that one major hydrological resource, many implications flow. The increase in surface water volumes and humidity has already ramped up the frequency of mudslides that knock out irrigation canals and bury fields. The retreat of the glaciers potentially affects wild onions that bloom in the icy snowmelt just below the lip of the ice masses where glaciers and lakes carve basins known as cirques. The instability of crop and wild plant yields could potentially disrupt the level of food security that workers on the land have traditionally maintained through the diversity of Pamiri fields and orchards. Any resulting diminishment of food security may undermine the health and welfare of the mountain-loving Pamiri peoples; and staple crops may never reach the cities on the plains below the Pamirs, which are just as dependent on them as the highland villages are. Ultimately, it may erode the food security of all of us whose diets draw on the food crops that originate in that hotspot of agricultural biodiversity.

Much of Vavilov's attention in the Pamirs was focused on determining just how high up on the so-called roof of the world various traditional crops could grow. He took meticulous notes on the elevational gradients over which he found nine different crops cultivated by the Badakhshani farmers. An Achilles' heel of Russian agriculture was its vulnerability to cold, dry weather, and he surmised that hardy, quick-maturing crops developed high in the Pamirs might be successfully grown in the northernmost reaches of Russia's grain belt. As a matter of course, Vavilov carried with him field instruments to determine the highest altitude that each grain variety, vegetable, and fruit could be grown, hoping it might give him a basis for selecting seeds to take north with him some twelve hundred miles to Saint Petersburg.

It is this set of observations that Vavilov made in September of 1916 that I had come to Tajikistan to redo some ninety years later. To help me interpret

Dr. Ogonazar Aknazarov and Pamiri potato harvesters near the River Panj.

any changes we might document, I had the good fortune to have as a collaborator the region's most accomplished botanist, Dr. Ogonazar Aknazarov, a local Pamiri, an internationally renowned scientist, and director of the Pamir Biological Institute. Ogonazar, a graying but still physically active botanist, can identify virtually every wild plant and crop variety in the Pamirs by sight, taste, and fragrance. For the last several years, he and his team have been engaged in efforts to document how global climatic destabilization is affecting the flora and vegetation of the region. His ecological understanding of the responses of wild plants to shifts in weather patterns at one of the world's highest botanical gardens made him an ideal team leader for those of us puzzling over the responses of crop plants to the same factors.

"Why the Pamirs are of so much interest to ecologists from all parts of the world," he explained to me one day while we were climbing into the Khuf Valley, "is because of these unbelievably steep gradients." He held his hand out the window of our Russian-built four-wheel drive and let his fingers soar like the wing of a hawk. We had started at a little below six thousand feet on the River Panj just four miles back, but we would be topping out at over ten thousand feet by the time we reached the higher edges of the village of Khuf four miles before us.

As our van lunged in low gear up a formidably rocky road, Ogonazar stuck his head out the window and recited to us the names of plants in Tajik, Russian, and Latin. Soon our driver shook his head and told us that the vehicle was overheating from the struggle up the slopes. He pulled it over to the side of the gravel road and encouraged us to walk ahead; he would catch up with us once he had added cool water to the radiator.

The slopes on either side of us were angling up sixty to eighty degrees above the road and dropping off several hundred feet below. The ridges were largely barren except for occasional patches of greenery where a leaky irrigation canal edged along its flanks. But as we wound up one switchback after another and turned a corner where the landscape opened out before us, we were surprised to see two or three acres of solid greenery on a slope that initially looked as precipitous as the others around us. As we moved closer to this anomaly in the landscape, Ogonazar spotted two men working there and called out to them in Rushani, the local dialect. They waved back, beckoning us to join them.

They offered us tea and fresh fruit they had harvested within the hour. As they told us about the fruit, the older man mentioned that a few years earlier, he had planted the very patch where we sat as a means to determine what fruit varieties could still produce yields near his home ground. A Johnny Appleseed of sorts, he had planted a mixture of fruit tree varieties at numerous elevations both above and below Khuf village, hoping that some would produce better than what the villagers were currently obtaining from their orchards.

As we talked to these men and to others in the Khuf Valley, one concern was echoed over and over again: They were seeking additional options in food production, because what had worked in the past for their forebears was not necessarily working today. They had found little value in the many so-called high-yielding varieties that had been distributed to them during the Soviet era and were just as skeptical of some of the newer varieties that had arrived from Western Europe and the United States since perestroika. Those "widely adapted" crop varieties simply hadn't performed well under their challenging climatic conditions.

When we asked whether their local climate was getting warmer, their answers, at first, seemed puzzling: "It depends where you are," the Rushani-speaking Johnny Appleseed replied. "At the very bottom of the valley, where our stream flows into the Panj, it is far colder in the summer than it was when we were children. We used to sleep outside in the summer there, but now it is too chilly at night. Some crops mature more slowly there than they used to."

Paradoxically, in the higher stretches of the valley above Khuf, temperatures were getting much hotter and the growing season was lengthening. As the glaciers melted more rapidly, less cold air stayed at the top of the valleys below the ice mass; the growing season there had lengthened by at least two weeks. Adjusting to this destabilization of their climate, local farmers had begun to sow wheat and certain vegetable crops at elevations that were much higher than where their grandparents had sown the same crops during Vavilov's first visit. At the same time, barley, rye, and oats—the traditional crops in the highest places—were being pushed out of their high-elevation fields by the expansion of wheat and potatoes. Furthermore, shepherds were now pasturing their Karakul sheep in alpine meadows above thirteen thousand feet, allowing them to compete for forage with the wild Marco Polo sheep, but also risking exposure to attacks by snow leopards. Recent changes in the climate thus appeared to be pitting one traditional seed crop against another and one form of domestic livestock against its wild kin. But farmers were also breeding new varieties.

The same week that Ogonazar and I ascended into the Khuf Valley, a colleague who now teaches at Cornell University—Dr. Karim-Aly Kassam—came out of the mountains on the Afghan side of the border to offer us a provisional summary of some 120 interviews he had done with farmers that summer. The highlights of his field investigations reinforced what we had been hearing, but, if anything, were even more disturbing. Apparently as the result of recent climate changes, many of the farmers with whom Karim-Aly had spoken were now planting as well as harvesting their annual grain crops two to four weeks earlier than they had done historically, and they, too, were planting some grains and vegetables higher than in the past. What alarmed Karim-Aly most was that traditional varieties of apricots and mulberries were so stressed that they were no longer producing fruit at the lower elevations where they had been grown for centuries.

Using Karim-Aly's field report as a lead, I returned to Vavilov's 1916 journals from the Pamirs, as well as the accounts of two earlier explorers—Olufsen and Korzinsky—who summered in the Pamirs in the 1890s, nearly two decades before Vavilov. Eight of the field crops whose observed maximum elevations Vavilov had recorded were now being grown an average of 1,387 feet higher than in 1916. When I compared our data with that recorded by Olufsen and Korzinsky in the 1890s, nine crops had shifted upward, as much as 1,561 and 1,661 feet in elevation. As the climate shifted, annual crops were being added to it.

It was clear from the historic accounts that during earlier eras, mulberries had been a major tree crop for Pamiri families. In fact, a typical Pamiri staple a century ago was the mulberry bread called *tut-pikht,* which Vavilov had relished. During my visit to the Pamirs, I could find families that still prepared tut-pikht on occasion, but not much of it was to be seen in households or markets that I frequented. The tree had grown rarer in the landscape and was no longer producing as much fruit.

A similar story lies hidden in apricot orchards. Some eight to ten decades prior to my visit to the Pamirs, apricots were the most widely planted fruit tree in the region, and wild apricots could be found in nearby orchards, as well. Although apricot trees persist in the region, we were told that erratic weather has drastically reduced their yields in many localities.

While sheep could climb or descend mountainsides with changes in the abundance of forage, and annual crops could be adapted to higher elevations each year, perennial crops such as apricots and mulberries did not offer such flexibility. It appears that they have instead been caught between a rock and a hard place—unable to fully fruit, but not yet stressed enough to die.

As Karim-Aly described the consequences of local weather patterns as they were expressed in various fruits, Ogonazar and I speculated on what processes might be at work. Both of us reached back into our training as horticulturalists and recalled that many species of fruit trees require exposure to a specific number of "chilling hours" during the winter, in order to produce fruit during the following summer. Through a physiological process called *vernalization*, trees exposed to a minimum number of hours below eight degrees centigrade are able to generate new flowers, which, if pollinated, ripen into edible fruit; trees that fail to meet such chill requirements are left barren. Has global warming at higher elevations around Khuf reduced the number of chill hours that traditional fruit varieties experience in winter so much that their minimum requirements are no longer being met?

Since we were novices in the landscape, Karim-Aly and I turned to our elder, Ogonazar, who had reflected on the triggering factors for these changes across his entire career. Although he had witnessed many unsettling patterns among both cultivated and wild plants, we found that he, too, had few firm answers. He was also wise enough not immediately to attribute all cropping shifts or stresses to physical factors alone:

> Yes, it may be true that global change is shifting the timing of harvests—it is now two weeks earlier in most places in the Pamirs for sure. Karim-Aly is correct that wheat is moving up in elevation, especially one particular local variety we call *safaiu dak*. But in the past, people not only sowed more barley because the growing season was shorter, but they were accustomed to making bread with it, as well. Due to Soviet influences, the peasants have become accustomed to think-

ing that wheat bread is more delicious than barley bread. At higher elevations, they have slowly shifted from sowing so much barley to sowing more wheat.

There is an additional reason the maximum elevation of wheat cultivation has climbed higher into the Pamirs over the last century, one that interested Vavilov a great deal. Beginning at least as early as the 1890s, the different dialect groups in the Panj watershed were actively trading their most prized seeds from valley to valley, and among them, several early-maturing wheats appeared. A few originally came from northern Afghanistan, on the edges of the Pamirs, but others were selected and named by Pamiri farmers themselves. The quicker their maturation times and the more tolerant they were to the peculiar spectrum of solar radiation where the atmosphere thins out, the higher in elevation they could be planted without suffering ill effects from that radiation or from shorter growing seasons. Thus, changing food preferences and local skills in seed selection must be factored in along with global climate change if we are to understand fully the last century of cropping shifts in the Pamirs, among other places.

Vavilov had been particularly excited when he met Abdul Nazarov, a local farmer who bred his own varieties and spoke both Russian and the local Shugni dialect. Nazarov alerted Vavilov to a locally adapted variety of wheat that matured a full twenty days before the others that grew in the region. Nazarov and his Afghan wife had obtained that cereal through one of those informal seed exchange networks that extends hundreds of miles southward all the way to Kabul.

In their local Shugni dialect, Abdul and his wife renamed their prized find *dzindham-dzhal-dak*, or "early-ripening wheat"; it is still found in the mountain villages where Nazarov descendants live today. Vavilov observed how successful Nazarov had been in introducing that wheat to so many of the isolated farming villages within the Pamirs—so much so that Nazarov had been left with hardly any seed of his own to offer the plant explorer. They finally convinced a few farmers to release just enough of their coveted wheat for Nikolay to take some back for study at his institute's fields outside Moscow. There what Nazarov and Vavilov suspected was confirmed: dzindham-dzhal-dak was the earliest-maturing wheat known at that time, and it was ideal for the high, dry climates of the north that had suffered countless famines due to the vagaries of Russian weather. Ninety years later, dzindham-dzhal-dak and its descendant varieties are once again allowing the Pamiri farmers to move their wheat production upward in elevation, offering them resilience in the face of climate change that they might not otherwise have.

While climate change is occurring everywhere—though perhaps not at the pace already evident in the Pamirs—the ways farmers are adapting to the changing conditions in the Western Pamirs are remarkable. The fruit tree planter we met in the Khuf Valley was not unique in his capacity for experimentation in the face of adversity. In nearly every one of the upland tributaries of the River Panj, a distinctive dialect group of Tajiks continue to select and name new varieties from their own crop gene pools while using their seed exchange networks to access new varieties from elsewhere. While they have been successful in keeping their annual crops moving upward to keep pace with climate change, their fruit trees are somewhat lagging behind, as any slower-growing perennial would. Yet these Johnny Appleseeds of the Pamirs are thinking ahead of the climate change curve and establishing orchards at places where it may take them another ten years to bear fruit.

As another man convinced us, he and his neighbors may be traditional farmers, but their practices are dynamic. He was the *mir ob*, or "ditch boss," of the Khuf Valley, a position of great honor. In each valley, there may be tens of kilometers of irrigation canals that meander along precipitous slopes, bringing water from just below the glaciers down to the fields surrounding a village. The task of maintaining those complex systems fell upon this middle-aged but lean, almost wiry man of many insights. He guided us high above the village, up through the highest sheep and goat pastures, until we could see where the glacial melt feeds directly into one of the irrigation canals. Huge boulders were strewn all around us; between them, we spotted the uppermost outpost for the herdsmen who brought their sheep up from Khuf village during the summer months. Just below a barren ridge, one hut with a high-pitched roof stood as a last resort for storm-stranded herders. It is by making observations along such a steep elevational gradient that Pamiri farmers and herdsmen accumulate traditional ecological knowledge to help them adapt to changing conditions. Lacking that, they would have to call for governmental or international "relief" as tens of thousands of less resilient farmers in the Tajikistan lowlands did following floods and landslides in the 1990s.

At last, the mir ob invited us into his hand-built house, where his wife served us homemade yogurt and sun-dried fruits from their previous season. Despite all the climatic changes and physical perils they have faced—and all the introductions of modern hybrids offered to them by governments and development agencies—the families of the Khuf Valley still favor their local varieties when they do bear fruit. While he showed us his seeds, one of the mir ob's elderly neighbors brought in some additional fruit, as well as an ancient stringed instrument. He then played a stirring ballad he had recently composed. I closed my eyes while listening to the song sung in dialect, savored the yogurt, and wished that Nikolay Vavilov could be there with us. He would have taken pleasure in the fact that a newly written ballad had been poured into an ancient form, one that still nourished and delighted those who lived and farmed on the roof of the world.

Of course, farmers in the Pamir highlands are not the only ones now facing a rate of climate change far more rapid than what occurred during Vavilov's lifetime. In the other countries where I would retrace his routes, I would also encounter evidence of climatic destabilization but confounded by human impacts: wasteful irrigation practices, salinization, groundwater depletion, surface water contamination, and over-allocation of surface water supplies. In other regions less remote than the Pamirs, farmers had been more inclined to accept the widely adapted hybrid seeds foisted upon them by corporations and agencies, at the expense of maintaining their finely adapted local varieties. That was one of many ways, as we shall see, that farmers have lost their resilience, and in doing so, have diminished their own food security over the long haul. Adapted seeds, of course, are not the only element that fosters food security and prevents famine. But as I would soon learn in visits to the Mediterranean countries where Vavilov once trekked, the diversity of a community's "seed portfolio" is in many ways one of the quickest indicators of its general agricultural health.

CHAPTER FOUR

Drought and the Decline of Variety: The Po Valley

Not long after retracing Vavilov's routes along the Silk Road in Central Asia, I had the occasion to visit one of the watersheds in southern Europe where the goods traded in from Central Asia and the Far East found their final resting place: the Po Valley, from Venice to Turin and Milan. I had some previous knowledge of the Po and its many flavorful fruits, so when I first read Nikolay Vavilov's commentaries on the agriculture of northern Italy, I unconsciously began to nod my head. He wrote,

> I have been to Italy a few times and crisscrossed it in all directions. I have visited Sicily [and] also studied Sardinia thoroughly. . . . But first of all I went to the granary of Italy—Lombardy—which extends from the foothills of the Italian Alps along the Po Valley. The richness of the deep and fertile soils has generated the impressive diversification of agricultural crops there. . . . In Italy, the high percentage of land under cultivation is striking. . . . [For this

> reason,] knowledge of Italy and its islands is of fundamental importance to any understanding of Mediterranean culture [and agriculture]. . . .

I, too, have crisscrossed Italy several times—once on foot for fourteen days, sauntering along from the chestnut groves in the mountains east of Florence to the vineyards, the bean fields, truffle forests, and olive orchards of Monte Subasio above Assisi. Yet, in order to grasp the changes that have occurred in Italian agriculture and crop diversity since Vavilov's first visit there in 1926, ten years after his first expedition to the Pamirs, I felt it would be best to return to the banks of the River Po, which forms Italy's longest agricultural valley. As the Po runs nearly west to east, from the Cottian Alps on the Italian border with France clear to the Adriatic, its watershed captures nearly a quarter of Italy. In Vavilov's era, the Po's agricultural valley, called the Val Padana, was the site of some of the most sophisticated and diverse irrigation agriculture in Europe. Vavilov was all too brief in his journal notes from Italy—perhaps because he was "honeymooning" there with Yelena, his second partner, after separating from Ekaterina—but it appears he was particularly impressed by the irrigated rice systems around Vercelli. The flood-irrigated rice paddies yielded as much as 8,000 kilos per hectare during his visit there, the farmers employing what he called "the peak of agricultural technology." While the Po's agricultural technologies still aid in feeding many more people than the watershed's sixteen million inhabitants, the environmental problems and damage associated with those technologies have become apparent since Vavilov's visits, leading *National Geographic* to claim that watershed-wide damage had turned the Po into "a river of pain and plenty."

Yet, exploring this valley today, it is not hard to be astonished by the abundance and antiquity of crops native to the Mediterranean that remain in place. Indeed, the Mediterranean Basin as a whole is its own hotspot of biodiversity, harboring some twenty 25,000 species of wild and weedy plants, 13,000 of them endemic or unique to the watersheds edging this "inland" sea. But what humbles me and other ethnobotanists whenever we are on the ground in northern Italy are the layers of cultural interaction with the flora that have been set down over many millennia, like layers of an onion. Vavilov capsulized the complexity of those interactions in two terse sentences: "A considerable portion of [Italy's] mountainous areas are covered with plantations trees planted in straight rows for fruit, nut or timber production whose trunks are entwined by grapevines, and whose interspaces between rows are seeded with wheat, fava beans, barley or other crops. The grapes here are 'wedded' to these trees. . . ."

The Italian reaches of the Mediterranean include annual crops domesticated in place, as well as crops transported here from other regions as many as six thousand years ago. There are other kinds of cultural interactions with the flora, as well: chestnut forests anciently tended for the sweet flour ground from their nuts, and oak forests managed for truffles and other "wild" mushrooms; weedy but delicious arugula and other herbs gleaned from trail sides; and vines laden with berries, capers, or grapes that are semi-managed in nearly every hedgerow. Perhaps the Italian palate appreciates wild and semi-domesticated flavors as much as it does domesticated tastes; perhaps more than most other European chefs, Italian cooks draw upon a wide range of food plants found in all stages of the domestication process.

One initial hint of the depth and complexity of the interaction between plants and people is the number of vernacular names for folk varieties and their crop species in local dialects. For the 551 species of cultivated plants that have been recorded in northern and central Italy, Italian farmers informally use no less than 10,672 vernacular names to refer to them. This must be some kind of world record for hyper-classification of agricultural resources, suggest crop geographer Karl Hammer and his Italian colleagues Pietro Perrino and Gaetano Laghetti. For instance, southern Italy and Sicily harbor about the same number of cultivated plants—521 species—but only 2,981 vernacular names for them have been recorded in that semiarid region. Korea also has roughly the same number of cultivated plants—578 species—but only 497 vernacular names have been recorded for them. Such elaborated classification of crop resources by northern Italians suggests how deeply they have been engaged in the recognition and selection of distinctive locally adapted varieties over the millennia.

It is not hard to notice that most rural Italians retain a passion for seasonally sampling the seemingly endless variation of vegetables and fruits with distinctive flavors and textures, for wines with different bouquets of fragrances, and for pastas made in a dizzying assortment of colors, shapes, and sizes. Nearly every country estate I passed through on foot near Pollenzo in the spring of 2007 sheltered a sizable and diverse vegetable garden; an orchard of peaches, apricots, figs, and now even kiwis; and a vineyard with several selections of grapes.

In the surrounding fields, several varieties of wheat were beginning to develop seed heads; some varieties were barbed, pale green, and multi-rowed, while others were barbless and darker green, with fewer rows of grain developing. In patches of woodlands remaining along hillsides and streams, it is hard to discern whether elderberries and chestnuts were intentionally planted or wild, though tolerated and tended. Of course, far downstream from where I hiked—where the Po Valley both widens and flattens—there is a larger scale of more intensive and homogeneous grain production, both of dry-farmed field corn, or *granoturco*, and irrigated paddies of rice. Both of those crops are now found on increasingly larger and more industrialized farms than those upstream in the Po's hinterlands, or those that Vavilov viewed eight decades earlier. But for all the production and prosperity wrested out of the Po in the past, the Italians that live within its reach are now realizing that this agricultural landscape may have reached the point of diminishing returns. As travel writer Erla Zwingle reported when she traversed the watershed in 2001,

> For everything the Po may have done for man, man has done at least as much to it. Nearly 25 percent of the land along its banks has been denuded of natural vegetation to make way for sterile plantations of poplars harvested for cellulose; the river is dammed for hydroelectric power and tainted by agricultural and industrial chemicals, to say nothing of the daily effluent from Milan. . . . More than half its total length is immured by man-made earthen embankments . . . which have only made the Po's floods fiercer and more disastrous.

By April of 2007, as I made my way down the valley from Torino toward Milano, the river of pain was no longer convulsing with floods; it was drying up. Droughts—exacerbated by wasteful irrigation technologies—had led to the demise of its once reliable productivity. It was plain to see that the wheat and the barley—and even the recently emerged corn plants—were stunted. Nearly all the local residents with whom I spoke worried that they would be suffering from crippled yields later in the year.

No wonder—they had just experienced the hottest, driest winter in two hundred years of weather records, accentuated by water wars between rural and urban sectors. The week before I arrived, so many stretches of the Po and its tributaries had dried up that the Italian government declared a state of emergency in the entire watershed, which normally generates a full third of the country's agricultural production. The ranges of the Alps situated in the headwaters of the Po had received some of the lightest snows in human memory, and the water reserves in the valley were reduced to a third of their normal volume at that time of year. In some stretches of the Po, the river levels were down six and a half meters below their long-running average during the spring season, when they are often swelling with floods generated by snowmelt.

When I spoke with Italian scientists and farmers, they acknowledged that while global climatic destabilization was taking its toll in their own backyards, drought was not the entire problem. True, experts had been predicting that average summer temperatures in the Po would be at least one degree centigrade higher than they had ever been. And if their predictions prove to be correct, not only will the tonnage of harvested grains, legumes, grapes, and fruits decline, but the electricity generated from hydropower will diminish, as well. Like farmers in many industrialized countries, farmers in Lombardy and Emiglia Romana have become accustomed to that cheap energy source for pumping water and doing other kinds of work. With Italian cities wishing to avoid blackouts, recently less hydropower has been made available for on-farm use. This coupling of declining water and energy available to farmers could potentially reduce the diversity of crops they can grow, for some cash crops such as sugar beets have formerly been pampered with lavish doses of irrigation water, tillage, and petrochemicals. A drought-induced crop failure might be worse in terms of its ultimate effects on food security, bringing some rare cultivated plant varieties to the brink of extinction. The effects of the 2007 drought in the Po may linger for years, as farmers attempt to recover from their earlier losses.

A significant part of the water woes in the Po have less to do with climate change per se than with outmoded technologies and policies for water conservation, however. That at least is the view of Ermete Realacci, who serves as chairman of the Italian Parliament's Committee on Environment, Territory and Public Works. "The real crisis is the lack of policy for the water sector," he said. For instance, orchard growers who use pressurized pipes and drip irrigation to grow fruit with a minimum of water waste are still paying seven and a half times what rice farmers pay, and the latter flood-irrigate their rice paddies by emptying the same open canals that Vavilov witnessed almost a century ago. Well before Vavilov was born, farmers along the Po had begun to use canal irrigation to grow more and more water-guzzling exotic crops—including rice, sugar beets, forage crops, tomatoes, and melons—in the valley. During his visit to the Lombardy region along the Po, Vavilov observed a "network of concrete-walled channels [that] conducts water to the fields. The walls of the channels, treated with vitriol, are free from algae and slime. The running water eliminates any traces of malaria-causing mosquitoes. Thanks to this experiment, Lombardy clearly demonstrates the possibility of growing rice in a healthy climate." Perhaps he could not have foreseen the water crisis that those rice paddies would precipitate decades later, but it could be that the concrete-lined canals he was taken to see were among the more efficient ones in the region at that time.

Today, most of the Po's industrial agriculturists are completely dependent on pumping water and chemicals through an eroding aqueduct system that was designed and built decades ago but never adequately maintained. Italy's environment minister, Alfonso Pecoraro Scanio, has admitted that the

Barefoot Berber harvester of dates,
Siwa oasis, Egypt.

Spring-fed oasis garden of Abou Shrouf, showing multiple strata of planting "hods" in waffle gardens of Siwa.

Badakhshani miller of stone-ground flour at water mill on Afghanistan-Tajikistan border.

Amharic-speaking girl presenting recently harvested teff, highland Ethiopia.

Blind Amharic-speaking farmer
harvesting teff, highland Ethiopia: instruments of the harvest.

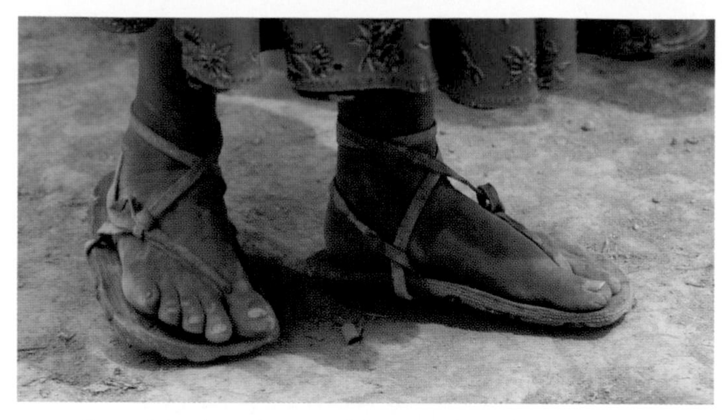

Feet of Tepehuan brewer of fermented corn beer, *tesguino,* in Pinos Altos, Chihuahua, Mexico.

Rarámuri (Tarahumara) farmer guiding horse-drawn plow, Sierra Tarahumara, Chihuahua, Mexico.

Rarámuri farmer hand-shelling blue maize, Sierra Tarahumara.

High elevation farming landscape in Khuf Valley of the Pamirs,
Gorno-Badakhshan, Tajikistan.

Shugni-speaking father and children harvesting grain in Gunt Valley of the Pamirs,
Gorno-Badakhshan autonomous province of Tajikistan.

Anciently cultivated olive grove in the Po Valley of Italy.

Bearded wheat fields of the Po Valley, with poppies in the foreground.

Fields of cabbages intermixed with pasturelands of Bekaa Valley,
beneath Mount Lebanon, in the southern cusp of the Fertile Crescent.

Ingano girls processing toxins out of cassava
in Rio Caqueta tributary of the Amazon in Colombia.

Typical farmhouse compound of highland Ethiopian teff farmers, with aloes and manure piles in foreground.

Patchwork of teff, barley, and wheat fields in highlands overlooking Blue Nile Gorge, Ethiopia.

current crisis in the Po would not be so severe if the infrastructure had been periodically repaired to conserve water, even in the best of years. "Our aqueducts are like sieves," he said. "Every year they lose 2,200,000 cubic meters of water at a cost of two and a half billion euros."

Tragically, there had already been considerable slippage, not just in water infrastructure but also in Italian agricultural traditions and the use of drought-hardy crops well before the effects of global climate change became so apparent. Karl Hammer and Gaetano Laghetti have calculated that since early in the twentieth century, Italy has been losing local varieties of cultivated plants as rapidly as the entire Mediterranean center of diversity has been losing wild plants. Roughly 90 percent of the diversity of locally adapted crop varieties of wheat has been lost in Italy since 1900. Of the seven species of wheat historically grown in Italy, five are now rare, and a sixth has not been seen in decades so is likely extinct in the countryside.

Fava beans have suffered dramatic declines in the number of native Italian varieties that remain in the landscape. The once great fruit diversity found in the Po has also been considerably eroded. Well-loved grape varieties have been pulled up over the last third of a century and replaced by field crops, and vineyard production in the Piedmont region of the Po has been cut by a third since 1970. Many folk varieties of vegetable crops such as *finoccio* or Florence fennel, which clearly originated in Italy, have gradually been replaced by just a few highly marketable cultivars. Roughly one additional local crop variety—long maintained by traditional farmers in the Po—is being lost in the region every second year. We cannot blame all of the losses on climate change; more likely than not, most are more strongly linked to the kind of wasteful capitalism that has worn down the irrigation infrastructure in the Val Padana and, to a large extent, depleted the water quality and quantity of the Po itself.

If there is a silver lining on the increasingly infrequent clouds floating over the Po, it is that this crisis has awakened many to take up concerns that should have been addressed long ago. In 2004, the rapidly growing international nonprofit Slow Food opened the doors of its Universitá degli Studi di

Scienze Gastronomiche (UNISG) in the heart of the Piedmont stretch of the Po Valley. Perhaps the first university to integrate agroecology and genetic conservation into the gastronomic sciences, UNISG has renovated a historic farm in the rural community of Pollenzo, not far from Torino and Milano. Some of the dramatic shifts in farming and food security occurring throughout the valley are readily apparent within a five-minute walk from the UNISG campus, so much so that they are a frequent subject of discussion among the international students who make Pollenzo their home for three years.

As a result of their mounting concern for what they call the "ecological emergency" of water shortages, 180 UNISG students traveled all 650 kilometers of the Po's watershed, from its headwaters to the Adriatic. Their twenty-five-day excursion by bicycle and boat was aimed at "Discovering the Great River," as the project was called, and assessing the state of its agricultural resources across four regions and thirteen provinces. The students not only have Vavilov's early notes as a benchmark from which to measure change, they will also compare their own impressions of contemporary farming with the detailed descriptions of the legendary Italian journalist Mario Soldati, who published *Travel in the Po Valley* exactly fifty years before the students embarked on their expedition. Vavilov envied northern Italian farmers for drawing on the highest level of agricultural technologies as well as making use of a high level of diversity in their locally adapted varieties, but both resources have suffered some erosion since his visits to the Po. Is it possible that the UNISG students' heightened awareness of the ecological emergency in the Po will motivate them to repair the damage done over the last century and restore Italy's place among the best guardians of diversified food systems in Europe?

In a sense, Slow Food founder Carlo Petrini is seeding a whole new generation of potential leaders for the rural communities of the Po. All of those seeds may not reach fruition, but the few that do may eventually help shape the future of the valley. My hunch is that after a month on foot in Italy's breadbasket, the students will never take their farming or food traditions for granted again.

There have been times in the history of Italy and of other countries when the entire agricultural fabric of a landscape has been unraveled by the ravages of outright war or by more subtle forms of imperialistic subjugation. While Italy has been largely free of warfare since the 1940s, that is not so for other Mediterranean countries where agriculture is even more ancient than in the Po. Vavilov had a prevailing curiosity about the entire Mediterranean basin, but one of the places where we can witness the most dramatic changes since the time of his travel is in Lebanon, on the southern tail of the Fertile Crescent—a place once touted as the original cradle of agricultural civilizations. There, the effects of colonization and war on agricultural diversity are as palpable as in any place on the face of this planet.

CHAPTER FIVE

From Breadbasket to Basket Case: The Levant

There were many reasons Nikolay Vavilov chose to seek out lands of the Middle East in the early summer of 1926, the same season he left Italy. In the nineteenth and twentieth centuries, the Fertile Crescent of the Middle East was considered *the* place where agriculture had begun. Vavilov's botanical and archaeological colleagues urged him to spend time in that presumed center of wheats, barley, and rye, and he was eager to do so. Nevertheless, he chose a tenuous time to visit.

He arrived in a politically fragmented landscape. The official partitioning of the region by French and English military leaders into the states of Lebanon, Damascus, and Aleppo had occurred less than six years earlier, in 1920, and it was still being contested and boycotted by most Muslim leaders. With considerable animosity emerging between the newly appointed officials of those various puppet states, it is not surprising that Vavilov struggled to obtain the papers needed to travel between the various states within what had been known as Greater Syria, or the Levant. Although his friend Mrs.

Vilmorin in Paris had helped him obtain the requisite visas, "on-paper permission" was not enough to guarantee him safe passage. When, at last, his Greek ship sailed into the harbor of Beirut in July of 1926, he suddenly realized that his problems had only just begun:

> It was difficult to select a less suitable time for an expedition to Syria. When I presented my passport with the French visa in the port of Beirut, it provoked great suspicion. . . . Later . . . when leaving Beirut . . . I came to understand the cause for the anxiety among the authorities. . . . All the mountainous lands south and east of Beirut were under martial law. Rebellious mountain tribes of the Druse had started a guerilla war against the French. . . . Our train required an armoured engine. . . .

My staff, too, wondered whether I could have chosen a less suitable time to plan a trip to Lebanon; in the "summer war" of 2006, the Israelis and Syrian-backed forces had used the southern half of Lebanon and much of Beirut as their stomping grounds, killing more civilians than soldiers. When I arrived eight months after the "end" of the war, most of the major bridges between Beirut and Damascus were badly damaged or altogether impassable, and drive-by assassinations were still common. The Lebanese economy was in ruins, and a new virulent rust disease that was attacking grain crops threatened to change the crop mix forever. Nevertheless, I had no trouble finding a great deal to be hopeful about in Lebanon's food and farming systems, amid all the obvious tragedy. I am not sure that Vavilov left the country with such an optimistic impression. What he did describe, however, was the tragedy that occurs whenever a country trades away its food security for export markets of cash crops, leaving it to gain most of its staples from beyond its own borders.

Vavilov abandoned the armored train that carried him out of Beirut for an automobile and some mules that he needed to get into areas reputed to harbor wild wheats, but he soon found that he was in more danger than ever. The authorities he met, unwilling to tell him where he might run into armed conflict, simply suggested that he hang a white flag from his automobile antenna to discourage both soldiers and guerilla combatants from shooting at him. Some errant shots at the vehicle occurred but harmed neither Vavilov nor his companions. If those incidents were not stressful enough, in the Lebanese lowlands Vavilov was bitten by mosquitoes that transmitted the malarial microbe, and his otherwise unflappable capacity to make fieldwork bear fruit under any condition withered on the vine: "Several bouts with malaria hampered my own work considerably. Instead of trying to collect the crumbling heads of wild wheat and wild barley under difficult circumstances, it was necessary to rest in bed several hours a day."

Vavilov's malaria made it a daunting task to collect as many wild wheats and barleys as he had hoped to gather in Lebanon, so he decided that he should cross the mountains into Syria, in the hope that Damascus would prove to be a more comfortable base for his recuperation. But as he moved eastward, he found himself increasingly excited and frustrated at the same time: "From there, we went to the oldest city in the world, famous Damascus [which] is full of gardens and surrounded by fertile fields. After completing long caravans lasting for days across the desert, the traveler enters Damascus and finds there a kind of 'El Dorado' with water and greenery. Unfortunately, Damascus was also under martial law [so] I had to limit myself to studying the grain market in the city itself and to visit only a few fields."

Eventually, he returned to the Bekaa, a stretch of the Fertile Crescent where wild wheat had recently been identified and promoted as a remarkable source of genes for breeding drought tolerance into bread wheat. Lacking the energy to explore more rugged terrain on foot in search of wild cereals, Vavilov mainly focused on some strains of wheat domesticated since ancient times that seemed to exhibit just as much resistance to drought as their wild relatives:

> The very first excursions to Arabian villages revealed a field which displayed wheats of a peculiar composition. Here I collected for the first time the singular subspecies which I later named "Khoranka" [locally called Hawrani]. This

> is a remarkably large-grained wheat with stiff straw and highly productive compact ears. [By the time these field notes were first published], the Khoranka had already been introduced onto tens of thousands of hectares in the highlands of Azerbiedjan. And right here [in the Bekaa]—on the slopes and at the edges of the fields—I saw for the first time stands of wild wheat. . . . But it was the drought resistance of the locally cultivated wheat, widely grown by the Arabian settlers, to which we gave our attention. . . .

The cultivated variety that Vavilov described near the village of Hawran is still grown to some extent, as are a few varieties such as Salamouni, which is ideally suited for making bulgur, a cracked cereal used in tabbouleh. However, while the locally adapted varieties suited to bulgur and another traditional dish, *kishk*, have persisted in Lebanon, bread wheats had largely been lost in the decades just prior to Vavilov's visit, when the Lebanese began to import their flour from Syria and Africa.

Still recovering from his illness, Vavilov had his crew help him gather and package seeds of such wheat varieties, while he recorded the habitats in which both cultivated species and their wild relatives grew. He also endeavored to make contacts with local scientists but found that the few well-educated ones had been sent to bureaucratic posts to help solve the severe economic problems of the region. He visited famous herbaria of cultivated plants but found them to be infested with insects and poorly maintained. He came away from Lebanon and Syria impressed by the prehistoric and historic wealth of the Great Syrian region but disturbed by the current state of its political affairs:

> Before leaving Syria, I went to Baalbek. . . . This country, with an ancient glorious culture that has extended through periods of impressive accomplishment, endures at present in an era of profound decline. It is under-populated and does not utilize its enormous natural resources to the extent possible, for they could support and provide opportunities for millions of people. . . . Nothing is left of the glorious past [for] during the ensuing dominion of the Ottoman Empire,

> the past fell into decay. The only highways are strategic military ones, constructed during the last couple of years, which uniquely indicates the influence of the French. . . . The conversion of Syria after World War I into a so-called French mandate in no way improved matters. In all of the large area of Syria—which exceeds in size even that of France itself—I found only a single agronomist. . . .

It was dawning on Vavilov that he had arrived in Lebanon in the absolute worst of times. Over the previous two decades, these lands had suffered from wars, locust plagues, economic disruptions, and out-migrations that had reduced the pre-1900 population by more than 60 percent. But the worst problem facing the Arabian farmers and herders whom Vavilov visited stemmed from their country's forfeiture of food security over the previous half century.

Beginning in the 1860s, when Napoleon III had landed six thousand troops on the Lebanese coast to intervene in internecine disputes on the side of Christians, the French had encouraged Lebanon's Maronite Christians of the highlands and valleys to abandon their subsistence crops in favor of growing mulberry trees for silk production. Spurred on by their own merchant class, the small shareholders in the Mount Lebanon highlands and the Bekaa Valley had planted nearly half of all their arable lands in mulberry trees, forsaking the wheat grains that had offered them bulgur for tabbouleh and the chickpeas from which they had made hummus over countless centuries.

As the number of silk mills in the region increased up through 1885, Lebanon began to gain half of its gross national product from the silk trade and was therefore able to purchase two-thirds of the bulgur, flour, dried legumes, and dairy products its peasants consumed from Syria, Turkey, and even North African countries. As Lebanese agricultural scientist Rami Zurayk has deftly summarized: "All this required the planting of large areas of mulberry trees, thereby displacing the traditional food farming systems, with the result that wheat and other foods had to be imported and were sold to the very farmers who had once grown them as staples."

With both the Ottoman War and World War I taking their tolls in the decade prior to Vavilov's visit, however, the silk-for-food trade network came unraveled. The resulting tragedy of food insecurity is the reason that some fifteen million people of Lebanese descent now live beyond Lebanon's borders, and only five million Lebanese remain in their native land.

In November of 1914, the Ottoman Empire entered World War I on the side of Germany, disrupting the trade of silk to France and leading the Ottomans to mandate that every Lebanese farmer who could carry a gun join the army. Hundreds of thousands of Lebanese men and boys were forced to leave the mulberry groves and remnant grain fields, though roughly a fifth of them deserted their compulsory military service and fled the country. The shortage of silk leaf harvesters alone would have thrown the country into an economic crisis, but blockades also prevented the women who worked in the silk factories from getting their products to high-end French markets, and the silk trade collapsed. With a locust plague devastating the remaining cereal fields in the summer of 1915 and no money to purchase staple foods from Syria or Africa, the food security of the Lebanese peasantry evaporated before their eyes. They were left with only locust-damaged mulberry leaves, which silkworms can eat but humans cannot. The mulberries themselves had little market, and their food value did not sustain many families.

In 1920, just six years before Vavilov set foot on Lebanese soil, a Maronite priest named Father Yammin described in horrifying detail the consequences of the loss of Lebanese food security in the summer of 1915. The locust plague was the least of it, for epidemics of typhus, cholera, and leprosy ravaged the working classes. Because many husbands either had been drafted or had fled the country and had no way to send money back to their wives, an unprecedented percentage of women were forced to work as prostitutes for the occupying forces of the Turks, and many succumbed to venereal disease or violence. The famine became so widespread in the countryside that some people survived only by scavenging meat from dead dogs, dead camels, or human corpses. By the end of World War I and the establishment of Lebanon as a sovereign nation in September 1920, 100,000 inhabitants of Beirut and

Mount Lebanon had died of famine or disease. In some agricultural villages of the Bekaa Valley, only one in twenty members of rural Arab families remained on the land. Amin al-Rihani, one of those who fled Lebanon during that diaspora, summed up the despondency and poverty that Vavilov witnessed: "Here are the ghost villages, inhabited by unemployment, laziness and desolation. . . . Here is the lost wealth, lamented by the newspapers. . . . National pride, dressed in artificial silk, eats its bread drenched in the sweat of Africa."

All this occurred in a country that had been one of the great breadbaskets of the earth, where many of the world's arid-adapted crops had originated, proliferated, diversified, and thrived. Vavilov arrived in the wake of that loss of food security, when one of the oldest farming valleys in the world had turned into a basket case. I take this story to heart, since my own grandparents fled Lebanon during that nightmarish era. I would probably be farming in the Fertile Crescent rather than writing this book if fate had not dealt the pitiful hand of cards that, among many others, the Nabhan family of Mount Lebanon and the Bekaa Valley received.

The loss of human life and viable livelihoods in Lebanon was accompanied by the loss of many time-tried local varieties of crops. No one knows for sure how many, or of what quality, because Vavilov arrived *after* the farm crisis of the years of the Ottoman War prior to partitioning, rather than before. But judging from the quality and uniqueness of the Khoranka wheat that Vavilov's crew gleaned while he was recovering from malaria, Lebanon once had seed of great importance, perhaps of heritage value, comparable to the architectural treasures still standing at Baalbek.

Some of those *baladi* (locally adapted) seed stocks undoubtedly did survive and can be found in a few patches where Lebanon's agricultural traditions hang on tenaciously today. They have somehow endured through Lebanon's era of agricultural modernization, which brought tractors and fertilizers to the Bekaa, displacing another 100,000 farmers and farm workers from 1955 to 1975. According to historian Fawwaz Traboulsi, just after World War II half of Lebanese families were still making their living off the land, raising field crops without the aid of greenhouses and livestock without the need for contained feedlots. But by 1975—just before another war devastated the Lebanese economy—only one-fifth of Lebanese remained involved in food production. Much of its traditional farming knowledge was lost as the population of Lebanon became predominantly urbanized, and the bulk of its food came from a dozen other countries around the world.

Having witnessed the state of Lebanon's food economy twice while visiting relatives in the Bekaa Valley, I was surprised to learn that in 2005, some recovery had begun. A group of farmers, activists, and food writers led by Kamal Mouzawak organized the first real farmers' market, featuring the traditional food producers remaining in Lebanon. Called *Souk el-Tayeb*—"the marketplace where goodness reigns"—the initiative began in downtown Beirut and has now spread to Byblos and Batroun. It is a place where urban consumers come face to face with the farmers, foragers, and bread makers who bring them the local cultivated variety of the melon cucumber called *me'ete*, the wild thyme known as *zaatar*, and the heavenly pastries stuffed with leafy greens from a cultivated purslane, known as *fatayer bi-baqleh*. According to Rami Zurayk's calculations, some sixty-five vendors at Souk el-Tayeb were already grossing a million dollars annually by the second full year that the market was open.

As I walked around Souk el-Tayeb one Saturday morning in the spring of 2007, I tallied some three dozen species of Lebanese-grown fruits and vegetables, in addition to spices and wheats sold in bulk, in specialty mixtures, and in freshly made artisanal pastries. No, not all of the vegetable varieties were grown from baladi seeds of Lebanese origin; some came from the finest French seed houses. And, yes, some of their fruits are now grown in high-input hothouses rather than out on the land, as they were traditionally grown. But Rami Zurayk and his colleagues have jump-started a "mobile agricultural clinic" that has assisted dozens of farmers with gaining organic certification and selecting varieties suited for elaborating value-added products. At the same time, Kamal Mouzawak has ensured that many of Lebanon's unique food folkways are featured in regional "Food and Feast" events that are now spreading to other parts of this little country.

Back in my own family's village of Kfar Zabad, on the eastern, arid edge of the Bekaa, I met with a dozen farmers—including some of my cousins—who have worked with the Society for the Protection of Nature to establish a special area known as a Hima, which uses traditional Muslim laws and practices to guide conservation and traditional use. They are among five dozen local farmers who are attempting to wean themselves from chemical fertilizers and high-input-requiring crops in order to protect the biodiversity of a globally important wetland for migratory birds.

According to Dr. Othman Llewellyn of Saudi Arabia's National Commission for Wildlife Conservation and Development, "The Hima is the most widespread and longstanding indigenous or traditional protected area institution in the Middle East, and perhaps on Earth. The Prophet Mohammad—upon him be blessings and peace—laid down general guidelines that transformed the Hima to become one of the essential instruments of conservation in Islamic law. He abolished the pre-Islamic practice of making private reserves for the exclusive use of powerful individuals."

By embracing the Hima concept as a tool for conserving the wild biodiversity of their local springs and wetlands, Muslim as well as Christian farmers in Kfar Zabad have decided to shift their farming practices to reduce contamination of the wetlands and to revive regionally unique specialty crops for sale to tourists attracted to the Hima. Whether they will go for organic certification or return to the baladi seeds of their ancestors has not yet been determined, but over sixty-five farming families have pledged to be involved in the comanagement of the Hima. Other communities in Lebanon and elsewhere in the Middle East have recently come to re-embrace the Hima conservation concept, restoring it to its historic position.

Such practices positively affecting land and water conservation have once again taken root in the Bekaa of Lebanon as well. As I talked with these farmers and saw the glimmer of hope in their eyes—despite the horrors of war they have recently survived—I felt blessed that I had arrived back in my grandfather's village at a most opportune time. Nikolay Vavilov would have been both gratified and relieved to know that traditional farming and other compatible land uses were being recovered, despite all the warfare this region had suffered.

Yet Lebanon at the start of the twenty-first century is a sober reminder that war is the worst enemy of food biodiversity and nutritional security. Few scientific reports from anywhere in the world have adequately documented how warfare devastates agro-biodiversity and security, but no firsthand observer could doubt that grave effects are evident in every battle-scarred landscape. As water resources become increasingly scarce and contested in the Middle East, as in other arid regions, we can unfortunately expect more devastation, not less. Community-based seed banks—perhaps with their collections backed up in more remote locations—are but one of many safeguards that must be employed to protect food diversity in the face of war. That the Souk el-Tayeb farmers' market remained active during the summer war of 2006 is particularly remarkable, but perhaps it offered hope, not just food, to those who retain allegiance to it. During such dark times, hope may indeed be a commodity as threatened as the seeds themselves.

CHAPTER SIX

Date Palm Oases and Desert Crops: The Maghreb

After Spain, Italy, and Lebanon, Vavilov set his sights on the southern Mediterranean, visiting Jordan, Palestine, Morocco, Algeria, Tunisia, and, briefly, Egypt. He knew that many of the crops important for future food security in the southern reaches of the Soviet Union had their ancestors and wild relatives in those arid and semiarid climes. But he was also painfully aware that many "alien and casual introductions" of plants and farming practices had already taken place in the Middle East and Maghreb of North Africa. Those modern introductions had likely obscured some of the history and eliminated some of the former food biodiversity in the region, but Vavilov remained eager to learn what of more ancient agricultural legacies persisted in Saharan desert oases in particular. However, in his 1926 journals collected in *Five Continents*, he lamented,

> An expedition to Egypt should have been next in turn, but endless attempts to obtain a visa did not produce any positive results. . . . Even the assistance given by

> Kurdali, the president of the Arabian Academy of Sciences in Damascus, led nowhere. So I engaged an intelligent Italian student, Gudzoni, to be my coworker. I prepared him and outfitted him with the necessary material . . . and sent him off to Egypt. Gudzoni carried out his mission [to collect locally adapted varieties along the Nile] conscientiously, while following the itinerary agreed upon through all the agricultural areas as far as to the Aswan area in Upper Egypt.

Although he was able to touch down in Alexandria later in 1926 and visit its souks and mosques, he was never able to gain the collaboration of Egypt's fine field scientists to go inland. Regardless, he still felt a need to inventory the food diversity associated with the date palm oases of North Africa's Maghreb region, which runs across the Sahara from Morocco to the Suez Canal, through both Arab and Berber lands. And so he set aside some travel time in 1926 and 1927 to sample crops in Morocco, Tunisia, and Algeria along the desert coast of the Mediterranean to Egypt's east. He hopped by boat from port to port—Casablanca, Algiers, and Tunis—before sailing around Egypt, down the Suez Canal, and on to Ethiopia. Some of North Africa's finest botanists and horticulturists met him in those ports; they replenished his food supplies and then accompanied him to the legendary oases of the Sahara strewn along the Spice Trail, the prehistoric trade route that had once lured Arab, Berber, and Jewish herb-seed brokers between Marrakesh and Cairo. From those wanderings, Vavilov surmised, "In general, North Africa is a kind of [ecological] unit unto its own. An analysis of plant geographic patterns clearly reveals a predominance of particular Mediterranean crops specific to the area. These are dominated by large-grained hard wheats [used in couscous] and six-rowed barleys that originated locally. The cultivation of large-seeded legumes and flax plants is concentrated along the coast as well."

Vavilov clearly had his eye out in those regions for wild relatives and primitive varieties of crops, using them as his primary means of discerning where crops may have originally been domesticated. But at another level, he was interested in the agricultural landscapes themselves and how they were traditionally managed. The date palm oases of the interior fascinated him the most, from both an agroecological and a cultural perspective. "Above all," he wrote, as if talking to himself, "it was necessary to get into the Sahara and see the oases. In July, [Louis] Trabut told me, 'only mad dogs and Englishmen' go there. But to find anything of the harvests, it was imperative to get there right away; to hesitate was out of the question."

His arrival at the first of the oases in July of 1926 could not have been better timed, for, he said,

> There was a whole forest of gigantic date palms already in fruit, which would [soon] ripen. . . . At the edges of the oasis there are Arabic buildings with flat roofs over which the date palms provide shade. There are small vegetable gardens full of carrots, beets and onions. The wheat, of course, was already harvested [but] going house to house I collected [woven wheat sheath] decorations made of seed heads which adorn the walls of the buildings.

Such oases, which range in size from a few hectares between dunes to hundreds of square kilometers, offer any traveler a stunning contrast to the surrounding sand sea of the Sahara, for they are packed with shady trees and understory crops. The mix of crop species of any oasis may already be familiar to most travelers, but particular locally adapted varieties are uniquely positioned in multitier plantings. Each of the Saharan oases I've visited has had its own character, but all share the soothing trickle of irrigation water coursing down canals, the breezy shelter of palm trees and bowers made from their fronds, the cackle of water-loving birds, and a year-round bounty of deliciously fresh vegetables. Perhaps because of the inherently startling contrast between the shady oases and the surrounding, open landscape of the "lifeless" desert, Vavilov remarked, "The entire oasis produces a strange impression. Deep furrows—through which water flows intermittently and periodically—have been provided for the irrigation of the trees. To walk there is always troublesome. The water lingers on the surface for several days [after each irrigation], keeping the ground soggy. From below the crown [of each date palm], enormous clusters of bright yellow fruit hang down, turning dark brown when ripened."

He was just as impressed by the herbs and vegetable gardens that are often found growing beneath the date palm canopies in the desert oases scattered along the Mediterranean coast, considering their size and quality to be exceptional:

> I came across enormous bulbs of extraordinary onions, weighing up to 2 kg. This was neither a coincidence nor something paradoxical. Beans, lentils, peas, wheat, barley, flax, wild carrots and wild vetch in this Mediterranean area are all distinguished by their unusual dimensions; their flowers, seeds and fruits are all gigantic, just as those of the onions are. The gigantism expressed in their [vegetative and reproductive] organs is a special morphological feature of plants in this Mediterranean region, as I later realized. Humans have undoubtedly played a role in [generating such gigantism], as has the intensity and antiquity of their agriculture. Nevertheless, natural selection has also favored the development and selection of . . . plant forms that seem gigantic compared to conventional fruits and vegetables, as their varieties of wild carrots and vetches demonstrate.

Vavilov attributed the enormous sizes of the flowers, fruits, and bulbs he encountered to abundant moisture and nutrients characteristic of the soils underlying the spring-fed oases. He was equally impressed by the distinctiveness of the cereal grains he collected, noting with awe that "the nature of the oasis has put its imprint upon them."

Eighty years later, in the heat of August, my own attempts to visit the oases of the Sahara landed me near the Egypt-Libya border but no farther west than that. Yet the Berber and Bedouin oasis that I studied most intensively was much like those that Vavilov visited in Morocco, Tunisia, and Algeria and of the same ilk of those his coworker Gudzoni sampled in Egypt. Most fortuitously, a contemporary of Vavilov, Arizona-based plant explorer Robert Humphrey Forbes, left detailed photographic and written records of the same oasis, Siwa, from an extended stay in 1919. Although Vavilov and Forbes apparently crossed paths only once—in 1930 at a lecture in Tucson, Arizona—their methodologies and goals were much the same: to seek out seeds and identify the adaptations of a diversity of crop varieties for evaluation and introduction into analogous environments. Thus, the benchmark data collected at the Siwa oasis by Forbes can serve as a surrogate for what Vavilov himself might have collected there if he had been allowed into the interior of Egypt.

During the era when Vavilov and Forbes wandered in the Sahara, it took several days by car and camel to reach Wahat Siwa, the Pearl of the Desert, from the Egyptian port town of Marsa Matruh. Wahat Siwa—the Siwan Depression—sits below sea level on the northern edge of the Great Sand Sea, some 315 kilometers south of the Mediterranean shores and less than 100 kilometers from the present-day Libyan border. After hours of crossing the barren plains, where only a few Awlad Ali Bedouins and their camels wander amid the cobbles and gravel, the wondrous sight of a million date palms and dozens of lakes appears to be a mirage. Suddenly, deep greens and blues replace the bleached-out tans and grays of the stony desert. The sharp aroma of oily, bitter desert herbs is replaced with the sweeter fragrance of orange blossoms, dates, mints, and hibiscus.

It is not a mirage; some thousand artesian springs feed into Siwa's lakes and ponds, two hundred of those springs directly irrigating some 3,800 hectares of date palms, fruit trees, gardens, and grain fields. The twenty thousand inhabitants of Siwa represent the largest human settlement for five hundred kilometers in any direction other than that of Marsa Matruh.

The oasis of Siwa leaves a strange impression, one perhaps different and more paradoxical than the impression Vavilov felt in oases farther west. There is at once a sense of uninterrupted continuity with the Berber oases of antiquity, with the mud walls of the Shali village compound rising high above thousands of palms, and a sense of rapid change, with tourist buses and European-style resorts evident all around the margins of that ancient compound. Since 1986, when the first paved bitumen road connected Siwa with the market economies of the rest of the world, the population of Siwa has more than doubled. Many of the new residents are neither Tasiwit-speaking Berbers nor Awlad Ali Badawi Bedouins but Arabs from Cairo or second-home Europeans who are economically engaged in making Siwa a great cultural and natural attraction.

While a million palms still cover the soggy, alkaline ground of the Siwan Depression, there have been some notable changes in what is grown beneath their canopies in the shady understory. The meticulous field notes of Robert Forbes from 1919, C. Dalrymple Belgrave from 1924, and Ahmed Fakhry from 1968 helped me evaluate changes in agro-biodiversity that had occurred between Vavilov's brief touchdown in Egypt in 1926 and my visits to the region between 2004 and 2006.

Date palms may still be the most prominent food crop that Siwans rely on for their own consumption and for export, but it appears that some changes in the varietal mix of date palms have occurred. Some historic reports claim that Siwan Berbers once grew dozens of folk varieties of dates, which their Awlad Ali Bedouin neighbors harvested and transported to Cairo and Alexandria. In 1832, travelers reported that as many as nine thousand camel loads of dates left Siwa for the Nile each season; a century and a half later, just before the paved road arrived in Siwa, the Bedouins had their camels carry ten thousand loads of dates across the desert to their traditional markets along the Nile. Today just five date varieties dominate Siwa's plantings, and the majority of the harvesting is done by migrant workers from the Upper Nile. Much of their harvest is transported by lorries or flat-bed trucks, and only two Siwan dates are regularly featured in the markets located elsewhere in Egypt and overseas: the flavorful world-class *sai'idi* date and the medicinal *tagtaggt*. Most of Siwa's other dates are much like the varieties grown in other oases, so they are less competitive in globalized markets. Fortunately, Slow Food International is assisting Siwans with the recovery of the rare varieties still found around the oasis and is helping market the entire range of Siwan date diversity in specialty shops.

Next to dates, olives have long been the second-most important perennial crop in Siwa. The Hamed olive from Siwa is world renowned, yet in recent years, European investors have introduced into Siwa many Kalamata trees from Greece and other cultivars from Spain. Because Egyptian labor costs far less than labor in southern Europe, those investors have attempted to undercut the prices for the same olives grown in Spain and Greece in the global marketplace. One set of investors from Canada, the United States, and the European Union has proposed increasing the number of olive trees grown in Siwa from seventy thousand to four million within the next decade, so that olive groves might eventually eclipse date plantations in economic importance. Those investors would no doubt plant more of the European cultivars than the time-tried Hamed native heirloom, unless forced by political pressures to do otherwise. Nevertheless, as of this writing, the perennial cover of Siwa offered by dates, olives, and jujube trees superficially looks much as it has looked for centuries.

What has already changed is the second tier of trees and shrubs grown at the oasis, and the vegetable varieties grown beneath those perennial fruits and nuts. Curiously, the number of species of fruits and nuts has increased at Siwa since the arrival of the paved road, because trucks can now carry in exotic nursery stock. Apples, guavas, prickly pear cactus, and bananas have been added to the traditional Mediterranean mix of fruits and nuts such as figs, pomegranates, apricots, peaches, mulberries, citrus, hibiscus, and grapes. Beneath those trees, shrubs, and vines, a diverse array of vegetables, grains, legumes, and spices are grown. European cucumbers have recently replaced the older snake melon cucumbers in dominance, and an heirloom called the honey melon has been replaced by modern cultivars of cantaloupe. Nevertheless, some thirty-three crop species recorded at Siwa during the era of plant explorations by Forbes, Gudzoni, and Vavilov remain in cultivation within the oasis, and many of the particular local varieties known as biladi heirlooms remain in Siwan cuisine.

One key factor fostering their in situ conservation is that the traditional mix of crops grown within a multitiered oasis garden offers Siwans a measure of resilience that industrial monoculture of grains or even of olives could never offer. Under the shade of dates grow peaches or apricot trees, which may have hibiscus or grapes beneath them, and onions or alfalfa below those. In years of drought, when flowing water ebbs and soil alkalinity rises, Siwan farmers can plant fewer of the salt-sensitive annual crops and fall back on their more hardy perennials for food. In more favorable years, when drought, locust plagues, and hot spells are minimized, they can add in more diversity under the protective canopy of their dates. However, few vegetable varieties introduced from the United States or Europe can withstand the heat or soil alkalinity that Siwa's baladi vegetables routinely tolerate. Because of those special environmental constraints, most Siwan farmers and gardeners still rely on the heirloom varieties that have stood the test of time.

Finally, Siwans are extremely proud of their traditional cuisine, which relies heavily on the particular mix of flavors, colors, textures, and fragrances that their own baladi varieties offer. Their local hibiscus flowers are yellow, not red, and are aesthetically preferred over the red *karkadeeh* flowers used as a tea throughout the rest of Egypt. The red hibiscus is now transported to Siwa in large quantities, where it is sold for cheaper prices, but Siwans still prefer their yellow variety. In contrasting their own culture with the Cairene Arabic culture of the Nile, the Berber people of Siwa point out that their foods are both better adapted to their oasis and more healthful as a diet. As long as cultural communities value traditional foods for such reasons more than they value the lower price tags on imported foods, their culinary traditions will persist.

Globalization of the Siwan food economy certainly began some time ago, but it appears that the homogenization of Siwan horticulture with that of the rest of the Mediterranean has lagged far behind that of other desert oases. Indeed, when Vavilov visited French horticulturist Louis Trabut in Algiers in the summer of 1926, Trabut had already introduced economic plants to Algerian oases from nearly every other tropical and arid subtropical country in the world. Vavilov was disheartened, he said, for "[my] first impression is that there is very little of the real Africa left there. All around and wherever you look in Algeria, there is an exclusively international flora: beautiful

Peruvian philodendrons with split leaves; enormous thickets of Australian eucalyptus; acacias and casuarinas; citrus trees introduced from southeastern Asia; Mexican cacti and agaves planted as fences along the shores."

Today, along the Egyptian shores of the Mediterranean from Alexandria all the way to Marsa Matruh some three hundred kilometers to the west, this same globalization of cultivated flora is proceeding at a blinding pace, as tens of thousands of shorefront condominiums are being built for wealthy European and Saudi vacationers. Except for ancient heirloom varieties of figs, which have thousands of years of tenure there, most of the food and ornamental crops along the North African coast could be found along each arid subtropical coastline anywhere in the world. Most require more fresh water than the desert has to offer. In little time, the cosmopolitan weeds of the horticultural world will swarm in on Siwa, as well, attempting to rob it of its distinctiveness. It is too early to tell whether the Berber and Bedouin values still strong in Siwa will be enough to absorb the insults without transforming the Pearl of the Desert into something altogether different.

Nevertheless, the trends in Northern African food diversity cannot be deduced from Egypt alone; from Ethiopia to Mauritania and Morocco, both the patterns of cropping and the pressures on farming cultures vary tremendously. Neither Siwa nor Egypt as a whole, for that matter, can serve as the best indicator of how African food diversity is currently being reshaped. A much more representative and at the same time dramatic example of agricultural change on the continent as a whole is evident south of Egypt at the headwaters of the Nile—the Ethiopian highlands. That is where Vavilov journeyed next, after boarding a ship in the Suez Canal, and that is where this story continues.

CHAPTER SEVEN

Finding Food in Famine's Wake: Ethiopia

Nikolay Vavilov arrived in what was then known as Abyssinia just before Orthodox Christmas in December of 1926, less than a year after his strange benefactor, Vladimir Ilyich Lenin, had died. Before Stalin's bureaucracy began to exert new pressures upon him, Vavilov worked with relative freedom, although his first request to undertake a major expedition was initially rejected. Once he argued that undertaking another major expedition would "contribute to the prestige of the USSR," his request was granted, so Nikolay spent several months preparing to direct a rather large entourage through the Ethiopian highlands and on to Eritrea. In addition to scientists and translators, twelve other men—mostly bilingual Abyssinians—would accompany him.

Perhaps Ethiopia was Vavilov's most excellent adventure. Nikolay was certainly not the first European explorer to set foot in the country, but he was the first Russian biologist to travel there, and he did so by train and by muleback. While his expedition may not have been as outright dangerous as several others occurring around that time, he still had to travel with rifles, revolvers, and spears to protect his group from crocodiles and thieves; to overcome the panic of market vendors who feared that he had the "evil eye"; to escape from late-night encounters with leopards; and to recover from both malaria and typhus, the latter of which nearly put him in his coffin.

Though neither the first nor the most perilous, the trip was easily the most productive of scientific expeditions to Ethiopia up until its time, in terms of its success in gathering seeds for future selection and use, in generating ideas that might help his country or others achieve food security, and in awakening recognition of Ethiopia's unique biocultural heritage. Earlier explorers such as Pedro Paéz, Richard Burton, and John Speke had sought fame by being the first to describe the headwaters of the Blue or the White Nile, while others sought to rescue the legendary Ark of the Covenant from Ethiopia's Emperor Menelik and his lineage. However, the European, Russian, and American public found Vavilov's quest for unusual seeds in Abyssinia just as exciting, such that his remarkable "discoveries" there gave the Ethiopian highland region its reputation as one of the more distinctive centers of crop origin and diversification on the planet. Vavilov's expeditions were regularly covered by the Russian, European and American press, as well as being widely celebrated among the diplomatic and scientific corps stationed in Ethiopia. Perhaps most important, the attention gave Ethiopians the pride and the inspiration to undertake a far more lasting effort toward conserving crops in situ than anyone of Vavilov's generation could have imagined possible in any country.

Thus Vavilov's place in Ethiopian history has been less heroic and more catalytic compared to that of Sir Richard Burton and his contemporaries, who are regularly referred to by travel guides in Ethiopia as "opening up" certain regions to scientific discovery. In fact, Vavilov's work helped to inspire a later generation of Ethiopian agricultural scientists like Melaku Worede, conservation biologists like Tewolde Berhan, and thousands of Ethiopian families who stuck with their "farmer's varieties" after attempts to lure them away to high-input hybrids after the famine of the 1980s. To the bafflement of most agricultural scientists and many experts on hunger alleviation, the total acreage devoted to indigenous crops like teff has increased in Ethiopia since the devastating social and ecological effects of the 1980s famine abated. However, nearly 2.5 million Ethiopians abandoned their homes and farmlands during the drought, and despite food aid as well as hybrid crop seeds given for planting, tens of thousands died during that famine. For those survivors who chose to return to their fields, the most drought-adapted of their native grains gained renewed popularity. While the famine may have been triggered by drought, it was horribly aggravated by political struggles that kept available food out of the hands of many of the hungry. Fortunately, since adequate rainfall returned in the late 1980s, the acreage planted to adapted crops like teff has increased by more than a half million hectares.

Teff is the millet-like cereal used to make Ethiopia's national dish, *enjera*—an enormous, delicious crepe on which dozens of different vegetables, sauces, and condiments are placed. Over the long haul, Ethiopian farmers clearly gained more resilience in their crops by using a diverse mix of locally adapted teff varieties than by investing in hybrids seeds and fertilizers imported from other regions.

When, in 2006, I first visited the Ethiopian Institute of Biodiversity Conservation, which harbors one of the first great seed banks ever to be supported in a developing country, I was reminded of the inspiration that Ethiopian scientists and policy makers have gained from Vavilov's example. Ethiopia's interest in protecting its own seed heritage has been linked by some historians to an incident in the 1950s, when cross-bred barleys from Ethiopia saved the entire California barley crop from a yellow dwarf virus, allowing millions of dollars to be made in California without any substantive benefits returned to Ethiopia. To keep such events from becoming commonplace—perhaps at the expense of their own people's welfare—Ethiopian scientists and policy makers opted for more control of their own resources. Begun three decades before my visit and initially known as the Ethiopian Plant Genetic Resources Center, the institute now works to foster the conservation and local use of crop, livestock, medicinal, and microbial diversity. The institute's staff is acutely aware that it may not be politically correct for institutions in African countries to feature so prominently in their offices photos of "a great white explorer," but Vavilov's significance at the institute is not so much his own fieldwork as the affirmation that the work of the Ethiopian field scientists to keep their crops conserved in situ is of global importance.

As they walk you around the institute's hallways, the staff points out photos of Vavilov's own hand-drawn map of his route through Somalia, Abyssinia, and Eritrea—the latter two being uneasily bound together as Ethiopia—and of Vavilov famously posing with the young but already charismatic Abyssinian leader Ras Tafari (from whom the Rastafarians take their name), later known as Emperor Haile Selassie. Other photos on the wall were taken by Vavilov himself at the open-air markets in Addis Ababa, Harer, and Gonder, where he realized that hundreds of the crop varieties he was seeing—land races belonging to several dozen crop species—were known only in Ethiopia. "What is so interesting about the diversity of crops in Ethiopia is that so many are endemic, found *only* here," commented Melaku Worede, the Ethiopian agricultural scientist who formerly directed the Plant Genetic Resources Center. "And what makes Ethiopia so special that way? We have so much topographic diversity that somewhat isolates farmers in one place from those in other places, for we have a dissected landscape with a broad elevation range."

The dissected landscape not only fosters the isolation of crop varieties, but also isolates wild species. It might well be argued that Ethiopia is really two countries—or three if you include the periodically contested lands of Eritrea. The highlands of Ethiopia are now identified by biogeographers as the eastern Afro-montane center of diversity. The rest of Ethiopia is considered part of the broader Afro-tropical center of diversity, which encompasses the Horn of Africa; the latter

Dr. Melaku Worede, Ethiopia's renowned expert on agricultural biodiversity.

includes the coffee-growing lowlands of the south and east as well as the "bottom floor" of the Great Rift Valley. It is incredibly rich in the diversity of food crop species, whereas the highlands are more renowned for their diversity of locally adapted grain and legume varieties. Like Vavilov's expedition, my travels would be largely in the Ethiopian highlands, where roughly 5,200 species of wild plants have been found, 555 of them clearly identified as endemics. The Afro-montane region of Ethiopia also harbors 680 bird species and some 193 kinds of mammals as well.

In addition to the many factors that contribute to fostering the wild biodiversity in Ethiopia, conservation biologist Dr. Tewolde Berhan reminded me of a factor that has fostered the diversification of crops in the Ethiopian

Dr. Tewolde Birhan, Ethiopia's environment minister, and his wife, Dr. Sue Edwards, experts on highland Ethiopia's biodiversity.

highlands of the north: human culture over thousands of years and adaptations of plants to that long history.

Dr. Tewolde did not overlook the great diversity of more tropical crops, such as coffee, in the southern lowlands, where the wild flora has remained far more diverse than the anciently farmed highlands. He simply wanted to remind me that cultural influences on the Ethiopian flora indeed run deep, for some of the oldest remains of human existence have been found in the Great Rift Valley and on the coastal plains. Archaeologists suggest that cereal crop agriculture in Ethiopia may be fifteen thousand years old. That would suggest that this region became a center of domestication for crops some three to five thousand years earlier than similar regions on other continents.

From his extensive readings in physical and cultural geography, Vavilov deduced that this highly dissected, anciently inhabited land had something special to offer. More intuitively—or fortuitously perhaps—he selected a route from the lowlands to the highlands that afforded the best opportunity imaginable for collecting a remarkable range of samples of Ethiopia's crop diversity, especially of cereals, over a rather short period of time. From one field between Gonder and Aksum, he collected "thousands of ears" of a peculiar awnless hard wheat, which he called "a first class discovery." Hundreds of thousands of seed samples were collected by his expedition and then shipped back to Russia for evaluation and conservation. The progeny of some of those collections actually made it back to Ethiopia's seed bank decades later.

While Vavilov was out seed hunting, he realized that durum wheat had not evolved in Egypt as others had supposed. Instead, it had begun to diverge from other wheats in the Ethiopian highlands before spreading northward to Egypt and eastward to Oman. In fact, Vavilov's travels across Abyssinia made him rethink many of the geographic patterns that had been accepted by botanists since the de Candolles' era. It also made him more intent on physically mapping where he collected various seeds, so that other biogeographers might build on his work.

Worede later chuckled at the uncanny ability of Vavilov to pinpoint areas of high diversity as we looked at Vavilov's hand-annotated map together. "It is a curious fact that so much of the total diversity of Ethiopia's small grains can be found along the very route that Vavilov chose," he commented. "His seed samples—of cereals, at least—were remarkably representative of Ethiopia as a whole."

I decided that I should follow—to the extent that the current infrastructure of roads and bridges allowed—the same route that Vavilov took through a portion of the highlands between the Great Rift Valley and the Blue Nile Gorge. I invited gardener-photographer and longtime friend David Cavagnaro along for the ride. We would try to visit the very same grain, vegetable, and spice markets that Vavilov had been lured to some eight decades before us.

Vavilov's first impressions of the highlands are still helpful to any novice trying to shape a mental image of the vastness of this rich landscape; he came from the port of Djibouti, French Somalia, on the coast of the Red Sea, after making his way by steamer from Alexandria, Egypt, via the Suez Canal, and wrote the following:

> On the 27th of December of 1926, I rode on the train into the interior of Abyssinia . . . passing through the Somalian savanna with its sparse acacias, the

> train approached a mountainous area. Here, the steep climb began. The Abyssinian plateau rises above the Somalian savanna like a castle. The ascent became increasingly steep. Two locomotives were needed just to pull a few cars.
>
> [There] the primary agricultural area of Abyssinia is situated at altitudes between 1600 and 3000 meters. . . . The train stopped at the station of Dire Dawa, at a distance of about fifty kilometers from Harer, the first major rural center along my route. Although I had not yet reached Addis Ababa, I decided to stop there [in Harer and Dire Dawa] to begin my investigations. I did not know what lay ahead, nor how I would be received by the government. And so, with the help of acquaintances I had gotten to know during our voyage [down the Red Sea] together, I was able to organize a small caravan that in the course of a few days explored this impressive area for plant materials that we might collect.

Vavilov did not use the term *caravan* loosely, for he had brought more than fifty men with him as his entourage. A pack train of donkeys, horses, and other beasts of burden was purchased to carry food, gifts, firearms, plant presses, and photographic and meteorological equipment through a range of habitats, from freezing mountaintops to subtropical swamplands at the bottom of the Nile gorge. Ethiopian officials insisted that Vavilov take rifles, pistols, and chains—not simply to deal with any highwaymen who might try to rob him but also as a means of quelling any mutiny within the troupe of mule skinners and porters he had hired. Vavilov's journals lament that he was spending far too much time attempting to manage all the wage workers he had enlisted and too little time in making seed collections himself. Of course, sometimes he was communicating through three or four different languages—Russian to Italian to Amharic to some tribal dialect that had yet to be recognized by linguists, let alone written down. Exasperated by the chain of translators he had to employ, he began to study Amharic on his own in order to get closer to the traditional agricultural knowledge of the Abyssinians.

With that newly acquired skill, he made a beeline for the ripened fields of grain nestled into the slopes of the Great Rift Valley, rather than going first to the Abyssinian capital to gain letters of support from bureaucrats in Addis

Ababa. There in the high, dry countryside, he saw the many locally adapted grain varieties being harvested and threshed through the use of rather ancient techniques and technologies unlike those he had seen closer to Harer: "Everything seemed to be totally different here in the highlands, for all of the crop varieties turned out to be definitively endemic," he remarked. "I had happened to arrive at a most appropriate time. The crops were still standing; the harvest had just begun."

David Cavagnaro and I were just as fortunate, for the harvest of wheat, barley, lentils, and teff had begun just as we arrived in Ethiopia in January of 2006. In every direction we were driven from the capital city by our Ethiopian hosts we came upon patchworks of grain fields that were being harvested that very day. We visited with entire families and their work crews, who were singing as they hand-harvested their grain with sickles and scythes. Brahman oxen, horses, and donkeys circled around threshing floors where the wheat was separated from the chaff, and gangs of boys with pitchforks and shovels tossed the seeds high into the air, winnowing the chaff from the grain. Nearby, older women sat tending babies plopped down on tarps, scooping teff and wheat and barley and yellow-brown, green, and even orange-red lentils into burlap bags, pottery vessels, plastic jugs, and wicker baskets. The entire highlands seemed tinted with ambers, beiges, rusts, and pale yellows as field after field dried down enough to offer its yield to be gathered, gleaned, and cleaned by human hands.

Looking carefully at such harvest scenes—for they have scarcely changed in the years since he visited the highlands—Vavilov realized that each tawny field of ripened grain was not a homogeneous mass of a single cereal cultivar but a panoply of intermixed strains of grain that formed a resilient polyculture. He was intrigued; he wrote "that the fields display such an incredible mixture of varieties. It was necessary to collect hundreds of seedheads just to obtain a representative sample of the botanical composition of a single field."

Melaku Worede reminded me how many studies by Ethiopians themselves have confirmed Vavilov's hunch about on-farm variation: "As you know, farmers tend to mix seeds . . . they are great experimenters, great promoters of diversity. . . . [To understand how this diversity arose] we must take into account

Vavilov photo of Abyssinian wood plow between Fiche and Gonder, Ethiopia, 1927.

Vavilov photo of spice vendor at market near Debre Libanos, Ethiopia, 1927.

that the farmers' very methods of crop selection enhance landrace diversity because they have so many criteria that they are selecting for . . . maturation time, stalk height, seed color, flavor and texture, and so on. . . ."

David and I quickly became accustomed to the fact that few if any of the wheat or barley or teff fields we ambled through were likely to be monocultures. A recent field study of samples of teff from six northern and central regions of Ethiopia bears that out. Just as no two platters of vegetables and sauces on enjera are alike, no two teff fields are alike. At least, that's what researchers from the Debre Zeit Agricultural Research Center found when they tracked some fourteen morphological traits in the admixtures of teff strains found in particular fields across a range of elevations in six regions. They documented extremely wide variations of those traits within each field population they sampled—more variation, in fact, than had been presumed to exist between regions or between elevational zones. The field populations of teff were highly variable for traits such as grain and stalk weights, seed yield gleaned from main stalks versus side stalks, and number of days from planting to ripening. Such mixtures give teff populations considerable capacity to respond to varying conditions, from drought to cool, wet seasons; from winds to still weather; and from manure-fertilized to nutrient-limited soils.

Although he was—by training and inclination—a wheat and barley man, Vavilov fell in love with teff and with several other crops unique to Ethiopia, as well:

> For the first time, I saw some of the special endemic plants of Abyssinia . . . such as the peculiar grain called *teff*, a particular kind of small millet that produces a first-class flour used for flat-cakes in Abyssinia. . . . There was also a new oil-producing plant with black seeds called *ramtil* or *noog* . . . [and] special varieties, or perhaps species, of pepper grass, and a special, tall-growing safflower, as well as a unique late-ripening sesame. . . .

That oil-producing *noog* is also employed in the making of enjera.

Despite endless efforts by development agencies to get Ethiopians to grow grains other than teff, roughly fifty million Ethiopians still use enjera as their daily bread, consuming close to 1.6 million metric tons of teff flour a year. But hours before enjera cakes are put on the grill, the teff flour is mixed with water and a special set of spices. Next, that batter is set aside not merely to thicken, but to ferment, initiating a microbial process that adds both flavor and texture to the enjera cakes.

In each "field kitchen" where enjera is made, a local potter has shaped a circular clay *mit'ad* skillet an arm's length in diameter especially designed for grilling enjera, and the mit'ad sits upon a small aluminum-sided stove that burns fuelwood into a high-intensity fire. The wood is fed through a door in the front wall of the burner until the skillet is nearly red hot. When the cook believes it is hot enough, she splashes some water onto the disk to see if it sizzles at the proper intensity. If satisfied, she rubs the disk with a ragged cloth that had been soaked in a bowl of ground noog mustard oil, which she deftly whisks across the disk in a matter of seconds.

When the noog oil begins to sizzle, she uses a dipper cut from a gourd to scoop about two cups of frothy teff batter out of a two-liter vat. Quickly, she pours a ring of the batter onto the mit'ad at its outermost edges, then another ring ten centimeters or so wide within it, then another, until the entire disk is covered in a thin coat of bubbling batter. The enjera batter begins to solidify and rise from the mit'ad into a thick puffy crepe riddled with air pockets that look almost like the texture of tripe. The cook then covers the mit'ad with a lid of the same size that helps the top of the enjera cook to the same texture as the bottom. After two minutes, the cook removes the lid, lifts the edges of the enjera with a knife blade to see if they are done, and then grabs a flat coiled basket or woven plaque that she inserts under the enjera. In one rapid motion, she lifts the entire meter-wide crepe onto the basket and lets it cool for two or three minutes before adding it to a pile of enjera that looks for all the world like so many flapjacks on steroids.

While David and I could photograph and take notes on teff in both the field and the kitchen, many of the other endemic crops of Abyssinia, like noog, initially eluded us, until we were at last able to see the previous season's harvest piled high in the open-air markets. For several days, we played a game of geographic hopscotch, trying to retrace the path of Vavilov's caravan from market to market, from Ankober and Debre Birhan overlooking the Great Rift Valley on the eastern edge of the highlands, to Addis Ababa in its south-central stronghold, and northward through Debre Libanos and Fichè, where one crosses the Blue Nile Gorge on the way up to Gonder and the monumental ruins of the Axumite Empire.

Each market was different in scale, size, and antiquity; some now found themselves as sideshows to railroad tracks or paved highways, whereas others remained shows unto themselves, situated in the shade of the leafy canopies of tall trees. One open-air market looked for all the world like it belonged under the mythic "tree where man was born," for its vast and ancient canopy seemed to stretch out in every direction to shelter an entire community of traders and buyers who had come in from all reaches of the gorge.

The road to Ankober had been partially paved and its route straightened since Vavilov's era, but the town still teetered on a dissected volcanic ridge, as if one great wind might hurl it into the abyss of the Great Rift Valley. As Vavilov described it, it was "one of the former capitals of the country . . . a few hundred low houses, solidly constructed of stone, sat snugly on the ground."

Ankober was where Vavilov's caravan at last left the acacia-studded lowlands behind, topping out nearly three thousand meters above the bottom of the Great Rift. The peasants living from Ankober northward through Gonder were both Muslim and Christian. They conversed for the most part in Amharic—the dominant Semitic language of Ethiopia—rather than in one of the dozens of dialects of the Cushitic and Omotic languages spoken more in the south. The farming compounds that David and I visited in 2006 had remained as picturesque and as functional as when Vavilov had photographed them. The compounds consist of stone- or mud-walled farmhouses with conical thatched roofs, surrounded by gardens, manure piles, outbuildings, and living fences in turn surrounded by a patchwork quilt of fields and pastures.

This crazy quilt of cereals, legumes, pasture grasses, and vegetable patches draped down over slopes, some of them terraced by dry-laid basaltic cobble walls. Ethnobotanist Leah Samberg has documented as many as fifty different crop species being grown for food and forage along such gradients, from highland plateau to river bottom. Some ridges edging the highlands are unterraced and so unbelievably steep that any draft animals used to plow them had to be tethered to trees on the ridge tops to keep them falling into the abyss. Yet the ridges were covered with golden-stalked grains, gray-green lentils, fuzzy dark green fava beans, or dark brown earth that had been recently plowed and manured. These patches of earth were edged with fencerows of towering eucalyptus trees or hedges of arborescent aloes and proteas, salvias, prickly pears, tree euphorbias, and thistles. Within each farmhouse complex, there was inevitably a neatly stacked pile of cow manure patties used for fuel, a mound of hay shaped like a bread loaf, another of wheat and barley straw, a pen for livestock, and several storage sheds.

At last, we located the Ankober market that Vavilov had visited; it had moved at least two kilometers from a nearby ridge into the heart of town since his time and was now situated across from the bus depot. As it was closed for the holiday being celebrated as we arrived, David and I found an ecolodge, where we were each given a conical thatched-roof hut to sleep in. The lodge overlooked the original market site used through most of the early twentieth century. We were invited up the ridge from our huts to eat in a beautifully crafted longhouse, where we sampled the same honey wine and barley beer that Vavilov had shared with the headman of the village.

The next morning in Ankober, David and I wandered over to the new market, which comprised some thirty-six stands that sell all manner of fruits and vegetables every Tuesday and Saturday morning. The list of produce did not vary much from the inventory that Vavilov would have seen: durum wheat, teff, barley, corn, peas, favas, and lentils; sugar cane, peaches, bananas, oranges, melons, almonds, and lemons; chili peppers, tomatoes, cucumbers, pumpkins, green onions, carrots, lettuce, radish, potatoes, and tobacco. Although many of the species originated elsewhere, their landrace varieties were peculiarly localized.

We wondered where all the spices could be found but soon learned that they are often sold in separate specialty markets set up in the cities or at junctions along ancient routes where traders from different elevations and language families periodically congregate. The largest market at a major crossroads is in Addis Ababa, where hundreds of booths sprawl across many blocks on the western edge of the city amid shanty towns of recent refugees from the countryside. Vavilov found it to be "of enormous interest . . . [for] farmers arrived there early in the morning from all directions, bringing their grain to sale in shawls and bags which they spread out to market. . . . This kind of exhibition made it possible to survey within a short time what was cultivated in the country and what the rural inhabitants based their lives upon."

Still lacking much insight into the dried herbs and endemic oilseeds of the highlands, David and I traveled northward toward the Blue Nile Gorge. There, Vavilov spent several days traveling forty kilometers a day on a road connecting Muka Xurii, Debre Libanos, and Fichè, one that his men and donkeys found to be "tolerable, although merely a trail." North of there he ran into leopards, crocodiles, and hippopotamuses, and he began to see an increasing number of locally unique varieties of lentils, chickpeas, peas, and vetches as the mosaic of agricultural habitats became more heterogeneous. In recent years, some of the very same legume varieties that he first collected and described have been relocated by the staff of the Institute of Biodiversity Conservation or by the institute's collaborators from the organization now known as Bioversity International. Other legume varieties, particularly some peculiar forms of field peas, have never been encountered by another scientist since Vavilov's first expedition there. As commercial varieties have been introduced from Europe, North America, and Japan, many local varieties appear to have simply disappeared and may now be lost to humanity.

Debre Libanos, one of Vavilov's stopping places, was a pilgrimage point for Ethiopian Christians. There, around the sacred artesian springs flowing out of the cliffs overlooking the Blue Nile Gorge, local residents claim that a number of miracles have occurred over the centuries. It is also a place where gelada baboons cluster in large numbers on forest ledges above the gorge; according to the same locals, these herds will occasionally raid a homestead garden for its melons, bananas, and papayas. As we descended endless switchbacks from the plateau above the gorge, we could see the towers of the cathedral well ahead of us but did not catch a glimpse of the market until we were nearly on top of it. There can be little doubt that it has remained in the same spot where Vavilov found it, for it sprawled beneath a tree so tall, with such a spacious, shady canopy that it served as a refuge for any weary traveler passing its way. If it was not, in fact, "the tree where man was born," it certainly seemed like the tree where man and woman first traded seeds and healing herbs.

Beneath that canopy, there were no neatly framed vendors' stands as in Ankober; instead, a mob of herb traders showed their wares on shawls and blankets and baskets spread out before where they sat on the hard-packed clay. Each vendor sat in the midst of many multicolored piles, which he or she guarded from incidental "sampling" by visitors walking by.

Perhaps "walking" is not the appropriate word for how buyers had to maneuver among the many vendors; the vendors' wares were so thickly packed beneath the tree that one had to almost tiptoe between the displays so as not to stumble or fall into anyone's carefully sculpted conical piles of turmeric, chili powder, or cumin.

At last we saw the shiny black seeds of noog piled high, next to other oilseeds such as peppergrass and sesame. Next to the many colors of ground cumin and chili pepper were brilliant golden piles of ginger, as well as masses of intact, sinewy ginger roots. One woman hand-roasted and weighed various grades of Ethiopian wild and domesticated coffees right before our eyes, while another sold various colors and textures of sea salt nested in pale brown paper containers, each looking much like a scoop of vanilla ice cream in a sugar cone. There were countless medicinal herbs, as well as crystallized globules of Ethiopian myrrh and "false" frankincense. Surrounding all the herbalists were huge piles of blue Hubbard squash, papayas, onions, Swiss chard, kale, mustard greens, and pomelos.

I was reassured that we found such a timeless marketplace; of course, it now has vegetables from the New World as well as the Old, and plastic and fiberglass containers as well as pottery jars and baskets. Yet the cultural context of this vernacular trading was much the same as it had been for centuries. Regardless of the changes in the rest of the world, these traders had managed to maintain some modicum of continuity with their predecessors, sustaining one of the world's most celebrated cuisines. For the moment, regional tradition was holding its own against globalization. No Wal-Mart could offer what this one-tree, one-stop shopping place could offer as it stood watch over the tributaries of the Nile.

Back in Addis Ababa, I had the chance to ask our Ethiopian colleagues how they perceived the ways their country had changed since Vavilov's visit. Melaku Worede reflected in particular on changes since he founded the seed bank now integrated into the Institute of Biodiversity Conservation. Since that time, he has witnessed more grassroots seed conservation efforts through his role as international advisor to Seeds of Survival (SoS), which today is one of the larger nongovernment organizations in Africa dedicated to the conservation of on-farm agrobiodiversity through promoting farmers' varieties. He slowly made his way, leaning on a cane, into the garden to have tea with David and me. Once he began to speak, it was clear that he had a uniquely rich perspective on changes in Ethiopia's management of agrobiodiversity.

Early in his career, Dr. Melaku had accomplished genetic research using some of Vavilov's concepts to guide the search for unique plant material to use in crop improvement. When the International Plant Genetic Resources Institute—now Bioversity International—selected Ethiopia as the number one developing country where a world-class gene bank was needed, he was among those who facilitated financial support from the West German government to found the Ethiopian gene bank in 1976. He not only trained Ethiopian scientists in the lab and field work associated with genetic conservation, but also mobilized them during the drought of the 1980s, when threats to traditional agrobiodiversity loomed on the horizon of the Abyssinian highlands. At the same time, he and his former students fostered some new grassroots approaches through Seeds of Survival, pioneering the kind of farmer-centered conservation strategies that luminaries like Vandana Shiva, Winona LaDuke, and Kenny Ausubel began to promote decades later.

All of Dr. Melaku's previous work had perfectly positioned him to deal with the impact of the 1984–85 famine, which seriously threatened Ethiopia's remaining reserves of farmers' varieties of traditional seeds. A number of development agencies and multinational corporations used the drought and famine as apertures through which they introduced packages of high-yielding hybrid varieties, herbicides, and other technologies to replace local varieties requiring fewer inputs. For example, the Sasakawa Global 2000 development agency of Japan introduced hybrid maize varieties in a package with water-conserving mulches and the herbicides Lasso and Roundup. Unfortunately, that effort has been plagued with many of the same problems that caused the failure of earlier agricultural development efforts. As Seth Shames of Ecoagriculture Partners has demonstrated, few of the Ethiopian farmers offered the package could afford the added costs of the hybrid seed and the herbicides, which negated the additional income derived from modest increases in their corn yields. The dilemma posed by farmers lured into trying such packages during drought years is that by the time they realize that a high-yielding variety might cost them more than it is worth, they have abandoned the time-tried seeds of their local varieties.

As the famine triggered by drought, war, and politics began to ring the death knell for tens of thousands of Ethiopians, the Dergue government, which had replaced Haile Selassie, realized that its native crop resources were under great stress. It encouraged its Plant Genetic Resources Institute to collaborate with Seeds of Survival to conserve what they could of Ethiopia's unique food diversity. Rather than simply locking away rescued seeds in the institute's gene bank for later use, the collaborative effort invested in on-farm conservation and improvement of indigenous crops by the rural communities themselves. As Dr. Melaku affirmed to me and David, "We realized that we should do more conservation through on-farm use of these crop resources. But first we had to identify the values that guided the selection of seeds by farmers. . . . Farmers are very smart, they know what they want. . . .

We just had to understand their logic, their traditional means of community-based seed saving and exchange."

Dr. Melaku and his colleagues soon discovered that despite the seed scarcity created by drought-induced crop failures, many farmers had buried caches of seeds in the ground for future use, as their local traditions had guided them to do. Those as well as other farmers exchanged seeds up and down the steep elevational gradients of their region, which offered them more resilience in the face of a fluctuating climate. Rather than ignoring such traditional practices, Dr. Melaku and his students encouraged them, documenting their efficacy during times of stress. Working with the farmers who chose to remain on the land rather than fleeing to the cities for relief, SoS gradually developed an extensive network of on-farm seed conservation sites that eventually involved some thirty thousand rural families.

Perhaps the most interesting strategy devised within this era was planting in one field a diverse admixture of several farmers' varieties with different physiological tolerances and other adaptive responses among them. In almost all plots where such admixtures have been tracked across several years, their yields through time have been found to be higher and more stable than those of any single so-called high-yielding variety.

These admixtures or polycultures of various grains have recently taken on additional significance, for an extremely virulent strain of wheat stem rust has arrived in Ethiopia recently, after first being spotted in Uganda in 1999. The strain of black stem rust known as Ug99 has spread across East Africa, finding most wheat varieties grown in monoculture susceptible and leaving their sowings a withered, tangled mass of seedless stalks. The first line of defense proposed by pathologists from developed countries is a fungicide, but most poor farmers in Africa can afford neither the chemicals nor the equipment to employ them. Accordingly, *New Scientist* has reported that billions of people may be at nutritional risk should the "wheat super-blight" continue to spread; Nobel laureate wheat breeder Norman Borlaug conceded that "this thing has immense potential for social and human destruction."

Ethiopia's extant grain diversity may ultimately be a better line of defense against such rusts, as a team of Swedish and Ethiopian scientists documented with another virulent rust—leaf rust—that typically infests Ethiopian tetraploid wheats. The team found tremendous differences among landraces in severity of infection, survivorship, grain weight, and overall yield. Both partial resistance and tolerance to leaf rust were found in Ethiopian landraces, even though they had all formerly been dismissed as being equally vulnerable.

More important, perhaps, is that Ethiopian farmers seldom grow only wheat; they also grow barley and teff, which are not bothered by Ug99's virulence. By shifting the mix of grain species they grow—and, as they have done in recent years, increasing the plantings of teff relative to those of wheat—Ethiopian farmers can employ several viable strategies for resilience not found among monocultural wheat farmers. Of course, only time will tell whether they have *enough* resilience to reduce the probability and severity of another major famine, but such famines are never entirely the result of drought or plant disease.

The recent history of Ethiopia reminds us that social and political factors can override the cultural and agricultural resilience found in a multicultural society such as that in the highlands north of Addis Ababa. Even in Vavilov's time, saving diverse seeds may have been of paramount importance in providing a greater buffer against future famines, but any political regime change can radically affect society's capacity to equitably distribute available cereals and legumes so that it can feed itself. The tenacity shown by Ethiopian farmers and scientists in maintaining their diverse seed stocks is heroic, but it does not stem from a romantic impulse. They are well aware that outside political and economic pressures as well as internal struggles may arise again, potentially disrupting the stability of the Ethiopian food system. People must be prepared for such disruptions, and rely on their crop mixes as one of many strategies for maintaining resilience.

CHAPTER EIGHT

Apples and Boomtown Growth: Kazakhstan

The fragrance of the Kazakh forest was unlike any I have ever known, for the pervasive smell of ripening and rotting apples and pears filled my nostrils. At my feet, russet reds, blushing pinks, vibrant roses, and creamy yellows mottled the ground, where wildlife had half consumed many of the fruit that make this forest so bountiful. I had arrived in the place that was the ultimate source of the apples and pears I had eaten since childhood, a place I had tried to imagine since I first read about these "wild apple forests" while still a student many years ago. But the sight, taste, and smell of wild fruit was not my only thrill this particular day; I also had the good fortune to meet the Kazakh scientist who, more than any other man alive, deserved to be compared to the legendary Johnny Appleseed, and to Vavilov, as well.

I had always dreamed of meeting someone who had known Nikolay Ivanovich Vavilov—not someone who had merely shaken hands with him at a meeting, but some scientist or farm laborer who had worked with him in the field. That dream came true during a summer 2006 journey to Kazakhstan's most famous city, Almaty, formerly known as Alma-Ata. There, I found Dr. Aimak

Dzangaliev, one of the last students whom Vavilov had taken into his fold, in excellent health and still working in the wild apple forests of the *Tian Shan*, or Heavenly Mountains, just west of China, where he had first met Vavilov in 1929.

The story Dr. Dzangaliev offered not only gave me insight into Vavilov the person, but also made me fully aware that "the Vavilov legacy" includes the field work and vision of many other fine scientists, such as Dzangaliev. Professor Dzangaliev's work on the origins of apples has deservedly brought him world renown and praise from the likes of N. A. Nazarbaev, president of Kazakhstan; cider maker Frank Browning, author of *Apples: The Story of the Fruit of Temptation*; writer Michael Pollan, author of *The Botany of Desire*; and Joan Morgan, coauthor of *The Book of Apples*, who has tasted nearly every apple variety in the world. Dzangaliev had also played a pivotal role in protecting these wild apples from the urban expansion of Almaty, the biggest oil boomtown in all of Central Asia.

While my own travel to Almaty was virtually without incident, few travelers during Vavilov's time could make that claim. In July of 1929, Vavilov and his traveling companion at the time, botanist M. G. Popov, set out for the agricultural oases in China's Taklimakan desert, which lay in the rain shadow of the towering Tian Shan range, in Xinjiang Province. Those oases could only be reached overland from Alma-Ata by a caravan of horses, mules, and Bactrian camels, for Xinjiang Province—then called Chinese Turkmenistan—was at that time as remote as any place on the Eurasian continent. At one point, Vavilov and Popov split up, and Popov suffered an accident that landed him in the hospital; only later did he learn how Vavilov had nearly drowned crossing the Kyzyldarya River. As Vavilov later recorded in his journals, "My horse fell into a deep underwater hole, [ejecting me from the saddle] . . . to reach the riverbank, I had to swim with all my clothes on, while lugging my equipment. My instruments and possessions—including the anaeroid barometer, the camera, and my travel documents—were all badly damaged."

In September of 1929, as Vavilov was traversing a mountain pass in the Zailijskei Alatau above Alma-Ata, he lost two horses crossing an icy stretch of the trail. Left with only one other mount, he was eager to turn in this last overworked horse in Alma-Ata for a bit of a reprieve. Some time in a motorized vehicle had been promised him by Professor V. A. Dubyanskiy, who was there waiting for him. Their plan was to travel together by car to Jarkand—now known as Panfilova—and then make their way by whatever means available to the Chinese border.

What Vavilov and Dubyanskiy did not reckon on was the curiosity the Kazakhs had about any plant explorer who would come by horse all the way from Leningrad to their homeland. Soon word got around that a great scientist named Vavilov was going on with his expedition but had lost two horses and might be in need of new mounts. That word soon reached a stable owner on the edge of Alma-Ata, who decided to offer the expedition some of his finest steeds. He enlisted the fifteen-and-a-half-year-old Kazakh boy named Aimak to help bring the horses out to Vavilov's camp.

Nearly seventy years later—at the age of ninety-two—Aimak Dzangaliev had shrunken from the tall, robust figure who had once given cider maker Frank Browning a powerful handshake; when I met the old man, he was not much larger than most teenage boys. Still, he retained a remarkable intensity and energy, which he attributed to eating several kinds of wild apples every day. That intensity was evident as his large, almond-shaped eyes widened and his hands wildly gestured while he recounted what had happened next on Vavilov's journey through Kazakhstan.

Dr. Aimak Dzangaliev, Kazakhstan's expert on wild apple diversity.

To their surprise, when Dzangaliev and the stable owner offered the horses to Vavilov, he politely declined, noting that Dubyanskiy was bringing a motor vehicle out for them to use for a bit of local exploration before they set off for China. Then Vavilov looked at the boy, and asked the stable owner if young Dzangaliev could accompany him for a day while he checked out the nearby wild apple forests edging the slopes above Alma-Ata. Vavilov needed someone with him who spoke the local dialect of Kazakh and more or less knew the terrain. Aimak Dzangaliev grinned as he told of their time together: "Vavilov [visited the countryside with me] and reviewed everything around Almaty in just one day. . . . Given his genius, his mind figured out just about everything [regarding the ecology of the apples here]. . . ."

Vavilov's own notes, edited for inclusion in *Five Continents*, relate his side of the story:

> Alma-Ata literally means "Father of the Apple." Thickets of wild apples stretch out in every direction around the city as far as the slopes of the mountains, where they form substantial forests. In contrast to the small, wild apples of the Caucasian mountains, the wild apples of Kazakhstan have much larger fruit, ones that hardly differ in quality from the cultivated species.
>
> Because we arrived in September of 1929, we could examine the apples while they were reaching ripeness. We could see with our own eyes that this remarkable site was no less than a center of origin for the apple, for the domesticated forms did not measurably rank above the spontaneously growing ones in quality. In fact, it was rather difficult to even distinguish truly wild apples from those which had been cultivated. Some of the wild ecotypes in these forests were so good with respect to their quality and their size that they could have been taken directly from an orchard here without anyone knowing the difference. . . . Nevertheless, it should be noted that the orchards here already include some of the finest, largest and most flavorful of the European cultivars, not the least of which is the famous *Apport* [or Constantine] apple.

"*He figured out everything*," Aimak Dzangaliev repeated, as if all of his own research over the following half century had been anticipated by

Vavilov's notes from little more than a day in the field. He went on, recalling their meeting many decades before: "He had declined our horses, but before he went on to Jarkand, he devoted himself to our apples. . . . It became my dream to be with this learned man, this mind. As a person, I am usually calm and in control of myself, but this time, I was inspired. Why, I asked myself, have our wild apple trees attracted the attention of such a genius?"

Dzangaliev's answer to his own question set him on a trajectory that dominated the next seven decades of his life. If a world-renowned scholar from Leningrad risked his life to see the wild apple forests of Kazakhstan, weren't those forests worthy of further attention by the Kazakhs themselves? At age fifteen, Dzangaliev aspired to get enough education to be able to study with Vavilov, and then to return to Kazakhstan to study the wild fruits there as Vavilov himself would have done. Within another decade, Dzangaliev had made it to Leningrad to study with both Vavilov and his most accomplished collaborator, P. A. Zhukovsky, who continued Vavilov's team's research on the geographic origins of crops.

During the three years of Dzangaliev's studies that overlapped with Vavilov's tenure in Leningrad, the senior scientist was struggling for the survival of his institute but still periodically gave lectures to the entire student body and faculty. According to Dzangaliev,

> It was in a huge auditorium that Vavilov offered his lectures to the college. . . . The students were so eager for these events that they quickly filled up all the seats in the hall, leaving no choice for latecomers but to sit on the stairs. During those years, I heard Vavilov present his ideas several times, and even worked up the nerve to ask him a few questions, which he answered with tolerance and grace, so that all of us could understand his explanations. . . .
>
> "He was not only beautiful inside but outside as well," Dzangaliev recalled somberly, folding his hands before him on the table. He himself was simply dressed, in a dark smock-like shirt and black trousers. "Vavilov would wear

> beautiful vests, and was so elegant in his gestures and his speech. To me, Vavilov was like a czar, or perhaps, like a God. I dedicated myself to following his ways of going about doing this work, albeit under conditions different from those under which he himself worked."

By that time, Vavilov's capacity for fieldwork was legendary among the agricultural students. On their routes through Kazakhstan and Western China, Vavilov and Popov had collected and annotated some five hundred seed samples, taken a few thousand photos, and surveyed the wide range of environmental conditions where contemporary agriculture occurred. They also had visited prehistoric sites that had been tentatively linked to the origins of agriculture in the region, consulting with archaeologists about the seed remains found there. Extrapolating from such surveys in a variety of locales, and among a diversity of Central Asia's ethnic groups, Vavilov attempted to place the region into a larger evolutionary and agroecological context:

> The many environmental conditions found within such montane regions, each isolated by geographic barriers, generate the diversification of crop varieties across different agricultural settings. Broadly speaking, [much of] Asia continues to function as a living laboratory where one can see the evolutionary processes unfolding before one's own eyes; it is therefore possible to trace some of the most ancient agricultural civilizations back to their very roots. . . . As seen through the archaeological record and historic archives . . . the mountain cultures have diverged from one another in terms of their languages, habits and specific geographic adaptations, providing us with a foundation from which we can reconstruct the various stages in the process of crop evolution for those crops which still persist in Asia.

When Vavilov returned to Leningrad in November of 1929, he worked for another year and a half on a monograph entitled *The Wild Relatives of Fruit Trees of the Asian Part of the USSR and Caucasus, and the Problem of the Origin of Fruit Trees.* Three-quarters of a century later, Professor Dzangaliev; his wife, Tatiana Salova; and their friend P. M. Turekhanova completed the modern sequel of Vavilov's Central Asia survey. In it, they concluded that within Kazakhstan's flora of 6,000 species, there are at least 157 species that are either the direct precursors or close wild relatives of domesticated crops. They found that 90 percent of all cultivated fruits of the world's temperate zones had wild relatives or ancestors historically found in Kazakhstan's forests, confirming the forests' status—first suggested by Vavilov—as a center of origin for many of the planet's major fruit tree crops. According to Frank Browning, those forests contain roughly four times the genetic diversity in apples that has been found in all of the apples as yet sampled from Persia through central and northern Europe:

> Locked away in the genetic codes [of the apples in the ancient Asian forests] are the still unexplored possibilities of what an apple might become: Apples resistant to rots and blights and insects. Apples untouched by deep killing freezes. Apples of tantalizing and unknown tastes. Apples possessed of deep, rich skin tannins and tingling fresh fragrances that could be the basis of new untasted wines and ciders.

Vavilov found Kazakhstan's apple forests to be evocative of the Garden of Paradise. He knew, too, the forest imagery found in Dante, for he once wrote his second wife, Lyonchka, that "halfway along life's path, I strayed into a dark forest, a dense forest . . ."

When I first stepped into those forests with one of Dr. Dzangaliev's coworkers, I had to pinch myself. Rather than oaks or beeches or aspens or pines, decades-old trees loaded down with fully ripened apples and pears surrounded us. We were barely an hour beyond the city limits of Almaty and had crossed through dozens of commercial fruit orchards on the lower ridges edging the valley before we entered these steeply sloped, semi-managed forests. Our driver took us up a gravel road that crisscrossed a small stream, and we reached a locked gate through which only military vehicles could pass toward the Chinese border.

There, amid the other trees were a bewildering variety of apple-bearing trees and shrubs all belonging to the native species *Malus sieversii*, which Aimak Dzangaliev and Tatiana Salova had not only researched but had eaten day after day for decades. Tatiana still marvels at the diversity of forms that can be found within a single habitat:

> Look at them: There are apples from the size of a large marble to that of a small plum; some are very glossy, others are somewhat dull; their skins can be red, yellow green, or mottled red. . . . It is not surprising that when Vavilov came to Kazakhstan to look at plants he was so amazed. Nowhere else in the world do apples grow as a forest. That is one reason why he stated that this is probably where the apple was born, this was its birthing grounds.

Within the range of *Malus sieversii* in this region of Kazakhstan, Dzangaliev and Salova have catalogued more than fifty-six native forms of apples, twenty-six of which might be likened to wild ecotypes, with another thirty being natural or anciently semi-domesticated hybrids. Some are on the remaining wild edges of extensive valleys that have been intentionally planted with European cultivars of domesticated apples. Tatiana Salova finds this worrisome, because the balance has been tipped, with wild apple habitat declining and domesticated apple plantations increasing in area: "Only here can we find cultivated and wild trees crossing, but the high number of cultivated trees is now swamping the wild remnants. Dr. Dzangaliev and I are very worried that there are few places left anymore where wild trees grow nowadays without being surrounded by cultivated trees."

Some of the wild trees have been lost to the expansion of commercial apple orchards, but most have found their space usurped by urban expansion, with many places that formerly offered ideal conditions for the growth of apple trees now paved or built over. From a series of apple forest maps which Dzangaliev, Salova, and their colleagues have elaborated over the five decades, it is clear that between 70 to 80 percent of the apple forests in the mountains immediately surrounding Almaty have been lost since 1960. At that time, the human population of Almaty was about 456,000, but by the year 2000, it had more than doubled, reaching 1,140,000 inhabitants. Since 1964, the land area in high-density residential use within the Metro Almaty region has increased 125 percent, with condominiums and large hotels overrunning lands formerly lined with apple-tree plantings. Although Vavilov once wrote of Almaty that "all around the city, you could see a vast expanse of apples covering the foothills," Dzangaliev carried in his mind's eye a vision of just how much of that vast expanse had withered.

"It was bad enough that a million wild apple trees disappeared during the war," Dzangaliev said somberly, referring to World War II, when he himself lost several of his toes to frostbite. He sighed, and went on, wringing his long, beautiful hands:

> On the Chinese border near Jungar, the Soviet government used the apples to make vodka and jam, but then destroyed all the trees, burning them as firewood. As Kazakh people today, we are on the edge of another such abyss of genetic loss among our apples. That is why I have written a report to the Kazakh Commission of the Environment noting that less than thirty percent of the original stands of apples remain, and others will be lost if we don't do anything to protect them. I pointedly ask them, Do you want to destroy the tree shown on the national emblem of Kazakhstan?

But instead of accepting such losses and assuming that protecting forest remnants is enough, Dzangaliev has another future in mind for the apples of Kazakhstan. He has proposed forest restoration in the best remaining habitats, employing some twenty-seven clones of choice wild apples that would be transplanted back into selected niches fitting their ecological needs. But he is also putting these twenty-seven clones on the fast track for commercial production, making the assumption that most Kazakhs will never fully respect the rich genetic legacy of Kazakh apples if they do not gain some income and health benefits from them: "At one point, I took a step back, and asked myself what Vavilov would do with these apples if he were alive today. And so, in memory of Dr. Vavilov, I have dedicated myself to create new selections of apples for planting that are exclusively derived from wild materials."

It is an astonishing approach to linking the conservation of apples to both the health and economic well-being of the human communities in their center of origin and diversity. In essence, Dzangaliev is *bypassing* the narrowed-down gene pool of domesticated apples altogether, and selecting the most delicious and nutritious wild and semi-managed apples for future cultivation, natural hybridization and selection. "If you want good apples on your table," he smiled, thumping the tabletop in front of him, "we need to go back to the best characteristics in wild apples, the ones that are healthiest for humans. A widespread cultivar like Golden Delicious is simply not that good; it does not have good nutritional value nor much disease-resistance . . . We not only have to do better plant genetics work on these apples, but we have to fight against [reiterating] the errors of history."

Given his status as a novagenarian, it was a bit surprising to see how anxious Aimak Dzangaliev is to see some of these wild apples contribute to his country's food security once more. Perhaps it is because he painfully remembers the harsh winter weather of 1955, which killed off more than eight thousand hectares of Kazakhstan's fruit trees; and he knows from his own observations that a mix of the wild apple ecotypes can better withstand the stresses associated with a variable climate than a few big cultivars can.

"That's why I'm asking the Kazakh government for more money to get the production of these apples on the fast track before I die," he said. At a field station he helped establish in 1959 below Jungar Alatau, he is hoping to ramp up the regeneration of planting stock of wild apples in what has already become the largest nursery in Kazakhstan. He has also proposed a series of two to three square kilometer protected forests as wild apple reserves to be strategically placed around the city of Almaty. When I mentioned that his vision seemed quite ambitious, Dr. Dzangaliev's face suddenly broke into a broad smile.

"I have this joke with my wife, Tatiana. We say that when I go to heaven to see what happens after I die, St. Peter will meet me at the gates and ask me what I have done during my time on Earth that was good enough to allow my passage into heaven. I will reply that I created twenty-seven new varieties from wild apples, and with my wife, I helped create fourteen new varieties of apricots. St. Peter will then look astonished and ask me how many millions of people have already eaten them . . .

"'None yet, really,' I will answer, trying to explain to him that such research takes some time to bear fruit.

"'Well unless the people begin to eat your apples very soon,' St. Peter will tell me. 'You will be going to Hell.'

"And it is because of St. Peter," Aimak Dzangaliev said, laughing, "that I am anxious for my people to eat the apples of Kazakhstan once again."

As Dr. Dzangaliev left us in a taxi to return back home across the burgeoning city, I was left with an ironic image. I imagined this tiny-framed elder making his way through neighborhood after neighborhood where wild apples once grew, finding condominiums and apartment buildings taking their place. The smell of these neighborhoods that wafted through the evening air was no longer that of ripening apples and pears, but one of motor oil. A city which prided itself on being "the Fatherland of apples" had grown so large that it was literally squeezing some wild apple varieties out of existence.

In Kazakhstan, as in many other rapidly urbanizing countries around the world, farmlands are being permanently lost to subdivisions, shopping centers, and highway interchanges, while remaining wildland habitats are being cut up into smaller and smaller pieces, no longer able to support the fauna which pollinates flowers or disperses fruits. This trend would have disturbed Vavilov, for he had already begun to notice the resulting genetic erosion during his own lifetime. And yet a branch of science developed since Vavilov's era—*reconciliation ecology*—is attempting to determine the most successful means for keeping these trees, their pollinators, and their fruit dispersers in our midst. Reconciliation ecologists are not content with leaving future generations a few dysfunctional forests as postage stamp–sized museum pieces, but are working to ensure that such habitats remain healthy and their wildlife populations viable. Dzangaliev's own efforts to keep islands of apple habitat afloat in Almaty's urban sea are laudable, but it will take the additional work of another generation of Kazakh ecologists to keep the web of ecological relationships surrounding these apple trees fully functional.

CHAPTER NINE

Rediscovering America and Surviving the Dust Bowl: The U.S. Southwest

Despite the political dangers inherent in relationships between Russian Communists and American capitalists, Nikolay Vavilov had, over his career, become good friends with several American scientists, Harry Harlan and Homer Shantz among them. For a number of years, he had supervised and occasionally visited an office crew in New York City that had purchased thousands of packets of seeds commercially available from some two dozen American nurseries and vegetable catalogs for shipment, storage, evaluation, and use in the Soviet Union. By 1930—the year after his trip to Kazakhstan and two years after he had unsuccessfully attempted to resign as administrator to devote full time to field research and writing—Vavilov was ready for another visit to the United States. He not only wanted to see the American scientists he had previously hosted on his own home ground, but was eager to visit what he believed to be the last vestiges of ancient agricultural traditions left north of Mexico. At the same time, he was leaving Leningrad just as Stalin's collectivization of agricultural lands and labor was being touted as "the Great Break with the Past." Capitalistic Americans were eager to hear what Vavilov himself thought of the Great Break, which the Soviet bureaucracy claimed would bring a 35 percent increase in grain yields to their country. (Little did they know that within four year's time, the Great Break would lead to a Great Famine, one that killed at least five million and debilitated tens of millions of others in the countryside of the Soviet Union.)

On October 6, 1930, Vavilov created something of a stir in what was then the small desert town of Tucson, Arizona, when he lectured to a packed house at the University of Arizona Library Auditorium. Although many in the audience were wary of Russians and the spread of their peculiar form of socialism, they were also aware that Vavilov was the official guest of the university's president, Dr. Homer Leroy Shantz, who, like Vavilov, had spent years studying the flora and vegetation of Africa. Vavilov, dressed impeccably in a black suit, sat quietly while President Shantz welcomed him to Arizona. After Shantz's flowery introduction to his Russian colleague's work, Vavilov stepped to the podium and began to speak in Russian, introducing his topic, "The Origins of Cultivated Plants." As he watched the eyes roll in the audience—for some of those present certainly assumed that he was not fluent enough to present his lecture in English—he moved quickly from Russian to Persian to French to German to Italian and, finally, to English, repeating the title of his lecture in each language. He spoke with a crisp but not overpowering British accent for the rest of the evening, astounding the crowd, not merely with his extensive technical and vernacular vocabularies, but also with his worldliness, humor, and intellectual insights.

Perhaps even more unsettling was that Vavilov did not treat the United States as the center of the learned universe, nor as the primary breadbasket of the world. "It is really time to begin the discovery of America," he ceremoniously announced a few minutes into his lecture. He was referring to the numerous undescribed and totally neglected food species growing right at the back doorstep of the small land grant college that was hosting him.

"A *Russian* has the gall to come to our country and declare it undiscovered?" more than one skeptic must have whispered to the person sitting next to him. What is it that has yet to be discovered here, they wondered.

Vavilov proceeded to argue that Americans should be grateful to the farmers of developing countries, from which most of the food crops are ultimately derived. He offered the example of Abyssinia, where both he and Homer Shantz had extensively traveled after being befriended by Haile Selassie: "The world owes a great debt to Abyssinia," Vavilov declared. "Although it is one of the smallest countries, it is the home of the great majority of all species of wheat."

He went on to describe how most of the crops that humankind depended on for food security in the twentieth century were anciently domesticated in just six regions on five continents. Three-quarters of the plants grown in the United States today, he reminded his audience, had their origins in Africa, Asia, and Europe. He had endeavored to find the particular regions from which those crops originated, noting that "plants have a definite predilection for certain localities or geographic regions, such that their origins can be easily traced."

Well, then, the skeptics may have wondered, which of the world's major food crops originated in North America? Actually, very few have been documented, Vavilov undoubtedly replied, looking over to President Shantz, who concurred, nodding. But that did not mean that America lacked its own domesticated crops or was impoverished in its abundance of native edible plants. Just that day, President Shantz had accompanied Vavilov to the Tohono O'odham village of Sells, Arizona, where he had seen the native tepary bean and another desert plant that he believed to have been domesticated in Arizona. Just what plant might that be, the agricultural scientists in the room wondered.

"It is the devil's claw, or martynia," he said, and gestured for President Shantz to bring him a chain of the oddly shaped pods of the plant we now call *Proboscidea parviflora* variety *hohokamiana*. The crowd looked at the size of the "claws" on the pods that Vavilov held from his outstretched arm like a mobile—they were two to three times the size of the claws on the wild plants that grew around Tucson. Vavilov was before his time in identifying devil's claw as a uniquely North American crop; it has more recently been well documented as being cultivated among more than a dozen indigenous tribes in North America.

Vavilov also spoke of his interest in wild rice, sunflowers, Jerusalem artichokes, tobaccos, amaranths, and cranberries, which were already recognized as unique American contributions to horticulture. Since his previous visit to the United States, the list of crops recognized as having been domesticated in North America had grown to include maypops and maygrass, sumpweeds and chenopods, little barley and Sonoran panicgrass, Chickasaw plums and lowbush blueberries, among others. Yet few Americans of the era could name such crops, and far fewer had gardened or cooked with those cultigens.

The easiest means by which to make the Americans pay attention to their botanical riches, Vavilov surmised, was to make them jealous of the advances the Russians were making in researching and utilizing their own hidden treasures. He shared the latest news from his comrade Sergey Bukasov, who had encountered several species of semi-domesticated potatoes in the Latin American countries just south of Arizona. Vavilov boldly noted that his institute was interested in obtaining tubers of two wild potato species that grew on the edges of agricultural fields in Arizona and New Mexico: "With regard to potato [diversity], my institute has opened up a whole new world," he claimed. "It has shown that there are many 'untouched' species in Mexico and South America." Then he repeated his dramatic call: "It is really time to begin the discovery of America."

By this time, few in the audience doubted him any longer.

After Vavilov's lecture in Tucson, Dr. Shantz accompanied him on a trip north, with the ultimate destination of the Painted Desert, where Navajo and Hopi cultivators had mastered dry-farming and terrace gardening. They first drove up from Tucson, across the Gila River, and into the Valley of the Sun, where Phoenix was at the time an artificially lush oasis full of citrus, date palms, cotton, and dairy cows. Vavilov was intrigued by the economic possibilities of the giant saguaro cactus, but Shantz cautioned him that it was too slow growing to become a cultivated crop. Vavilov must have dismissed the caution, explaining that the Soviet Union had the patience to take on such long-term projects and had already achieved some positive results. Shantz later complained in his journal that the Soviets have "a wonderful capacity to confuse *plan* with *accomplishment*, and to conclude that what is decreed is already accomplished."

As they traveled, they rose in elevation through several "life zones," as Shantz and his ecologist colleagues called the various vegetation complexes associated with different elevations. They passed through some beautiful desert grasslands south of Ash Fork, Arizona; briefly stopped at the south rim of the Grand Canyon; then hurried on through the western edge of the Painted Desert to reach their lodging in Tuba City, Arizona, just before sun-

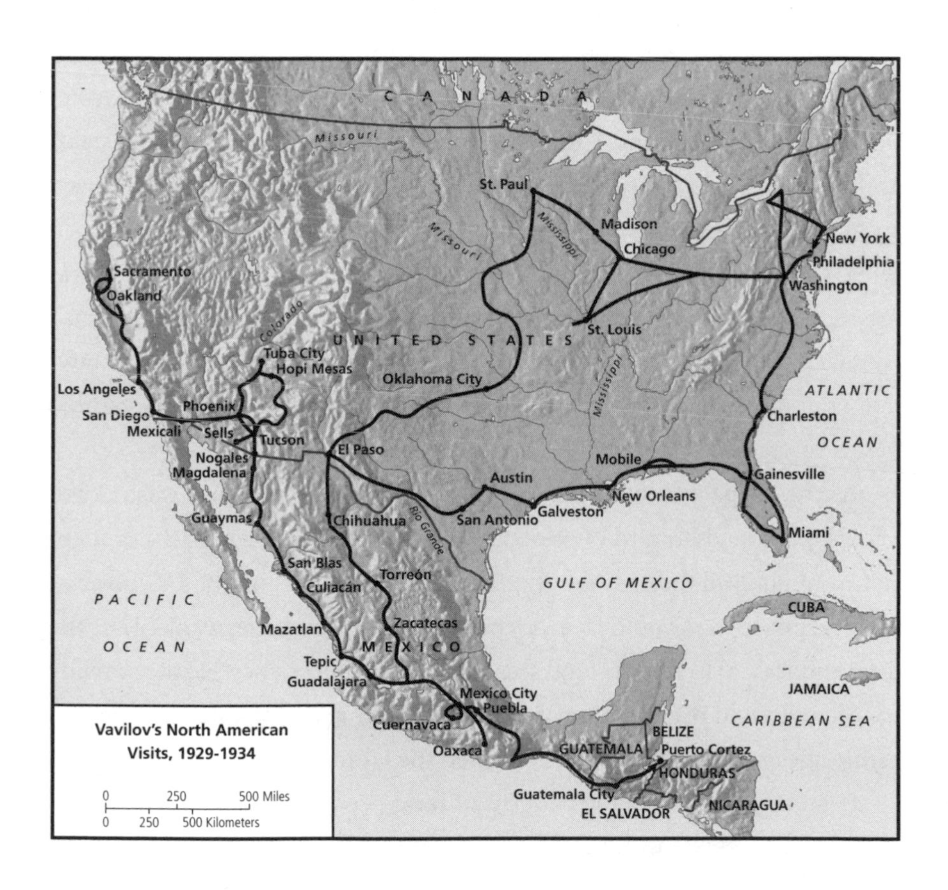

set. There in the rich golden light of an autumn dusk, they glimpsed abundant harvests being brought in by Hopi and Navajo families, even though the Dust Bowl drought was beginning to take a toll on the surrounding wild vegetation. Despite meager piñon nut harvests and minimal forage for the Navajo's sheep, both the irrigated and the dry-farmed fields had managed to yield considerable food that fall of 1930. Field notes taken that day by Shantz revealed the austerity that he and Vavilov felt surrounded the hard-working indigenous farmers: "Desert not an abstraction at this time—almost no flowers . . . [but] Hopi crops good—maize, melons, apples."

The next day, Shantz introduced Vavilov to Hopi farmers at Moenkopi, before they traveled westward fifty miles to Hotevilla, one of the ancient farming villages on the three Hopi mesas. It remains unclear whether Vavilov collected any corn, bean, devil's claw, squash, or other seeds from farmers during this short trip; the only "Indian corn" noted in his journals as being in his possession was given to him on an earlier trip and was from Wisconsin. Nevertheless, Vavilov and Shantz were astounded to see tepary beans and sunflowers being harvested from sand dune fields below mesas that received no moisture other than rainfall and runoff. They saw stacks of blue corn being piled high against the sides of stone walls; they also entered a *piki* house, where Hopi women poured a batter of blue cornmeal, water, and ash over a red-hot stone, producing wafer breads that reminded both Vavilov and Shantz of the enjera they had eaten in Ethiopia. The cultures of the Painted Desert impressed them as much as the landscape itself.

As they departed the Hopi mesas and followed the Oraibi wash down across Navajo lands toward Winslow, they also saw the first signs of the drought that would prevail throughout the 1930s. That drought and the U.S. government response to it would forever change the opportunities available to the Hopi and Navajo to achieve food security and self-sufficiency. Neither Vavilov nor Shantz could realize it then, but the devastating ecological and socioeconomic effects of the drought have affected the Hopi and Navajo to this day.

Let us try to imagine the patterns of food production and consumption on the Hopi and Navajo reservations in 1930. The summer rains had begun to fail; that year they provided most dry farms with hardly enough moisture to produce a yield in the fall equal to the amount of seed planted in the spring. The wild plants that typically abounded around springs and streams were being consumed by cattle and sheep that had no other forage resources to sustain them. Some Hopi and Navajo farmers and herders had begun to seek wage work off the reservation, as craftsmen and guides in nearby national parks.

Yet, even by 1936—at the height of the drought—some 2,779 Hopi persisted in planting the fields that Vavilov had visited, continuing to plant some 5,916 acres of land. During some of the toughest drought years Hopi farmers faced, they still managed to grow nearly 4 million pounds of corn and 57,000 pounds of beans, in addition to sizable quantities of melons, watermelons, squashes, pumpkins, peaches, apricots, pears, apples, grapes, and other garden vegetables. They grew roughly 99 percent of the maize they required for their own food consumption and produced more beans than they consumed. They sold their surplus crops to their neighbors, gaining more income from their corn sales during those years than from their sales of jewelry and basketry, or from their wild harvest of piñon nuts. Their annual purchase of food from outside sources was on the order of $68,000, or less than $25 per capita in 1936.

Their 1,949 Navajo neighbors around Tuba City farmed only 1,313 acres in 1936, for some of them were migratory sheepherders. Nevertheless, those who farmed managed to raise 1,800,000 pounds of corn, and 1,300 pounds of beans. During the height of the drought, when rainfall was less than 90 percent of the average, they also raised hay, potatoes, melons, watermelons, squashes, pumpkins, peaches, apricots, pears, apples, grapes, and other garden vegetables. They raised 96 percent of the corn they required for home consumption, but only 45 percent of the beans their families consumed. Their annual purchase of food was roughly $50 per capita.

In 1940, the drought subsided, but World War II flared up, taking many of the Hopi and Navajo men away from home to become soldiers and "code talkers." Even with some of their hardest-working men off to war, the families around Tuba City produced more corn than they could consume, selling a sur-

Hopi farmer in Arizona dry-farm cornfield visited by Shantz and Vavilov, 1930.

plus of 6,200 pounds of corn to traders. They bought only $42 of food per capita from traders, most of which (like coffee and sugar) was produced far away from their region. In short, the Hopi and Navajo of the Tuba City area continued to be 90 percent food self-sufficient up through the onset of World War II.

How was that possible for people who lived in a stretch of the Painted Desert that received less than ten inches of rainfall in the average year and far less during the Dust Bowl era? What buffered the Hopi and Navajo from famine during drought was a mixed subsistence strategy that drew on a diverse set of crops adapted to different agricultural habitats. Although the sand dune fields that Vavilov saw below the limestone mesas were the most obvious agricultural habitats, the Hopi and Navajo all had spring-fed terrace gardens hidden away in canyons on the "back sides" of the mesas. They also nested "floodwater fields" along ephemeral streams, where storm runoff could be harvested off the surrounding watershed. For each of those habitats, they had crop ecotypes finely adapted to the soil moisture conditions prevalent there; for instance, the maize varieties planted in sand dune fields could be planted fourteen inches deep where residual moisture could be found and emerge from such depths to produce a harvestable crop on minimal summer rainfall.

In 1935, ethnobotanist Alfred Whiting interviewed several Hopi farmers whose Moenkopi fields and gardens lie on the edge of Tuba City. Among just five families, Whiting found that the following crops were still grown on a regular basis: yellow, red, blue, white, violet, pink, and speckled flour and flint corns; purple-backed *kokoma* corn; sweet corn; white and gray lima beans; white and blue string beans; grease beans; pole beans; white tepary beans; peanuts; watermelons; casaba melons; honeydew melons; muskmelons; banana squash; cushaw squash; Hubbard squash; cucumbers; onions; chilies; tomatoes; turnips; red dye amaranths; cabbages; peaches; pears; apricots; apples; grapes; and cherries. No doubt, nonedible crops such as tobacco, gourds, and hay were also grown. Although the Hopi were already getting much of their vegetable seeds, onion sets, and fruit trees from nurseries off reservation, they were still growing most of their corn, beans, chilies, squashes, peaches, and apricots from their own seed.

Fifty-four years later, in 1989, with the permission of the Office of Hopi Lands, I interviewed descendants of some of the same families in Moenkopi that Whiting had interviewed, while my friends Daniela Soleri and David Cleveland interviewed other farming families on the three Hopi mesas. I found that the Hopi of Moenkopi were growing yellow, red, blue, white, pink, speckled flour and flint corns; purple-backed kokoma corn; sweet corn; white and gray lima beans; pole beans; white tepary beans; watermelons; casaba melons; honeydew melons; muskmelons; cushaw squash; Hubbard squash; cucumbers; onions, chilies; tomatoes; red dye amaranths; peaches; apricots; apples; grapes; and cherries. In addition and not noted by Whiting, they were growing almonds, pecans, walnuts, plums, carrots, cantaloupes, and "rotten" beans. A few crop varieties may have changed as more commercial seeds were brought in and evaluated, but many of the farmers still saved their own seeds of the varieties that proved to be productive in their irrigated fields and gardens.

When Daniela, David, and I discussed the similarities and differences between Whiting's surveys and our own a half century later, we could articulate a few prevailing patterns. As Daniela and David later wrote,

> Overall, Hopi farmers' crop repertoires are still dominated by traditional Hopi varieties. While experimentation with new varieties is constant, these experiments do not appear to be leading to the direct replacement of the landraces. This is especially true among the staple crops. . . . These findings suggest that those crops which play a central part of the Hopi subsistence strategy [and ceremonial life] are maintained through seed sources on the reservation. In comparison, crops which are more recent additions to the Hopi crop repertoire are more likely to be obtained from off-reservation sources.

You might presume that this time-lapse comparison may be enough to tell you all you need to know about changes in agrobiodiversity among the Hopi and Navajo since Vavilov's visit. But in 1997 another extended drought set in across the Painted Desert, and just about that time, Hopi and Navajo farmers realized that many of their perennial springs were drying up. It was difficult to discern whether the spring flows were diminishing due to drought alone, due to groundwater pumping for Peabody Energy's mining operations on Black Mesa above the Hopi villages, or due to both. However, it is clear that for the thirty-five years beginning in 1969, the Peabody Energy company pulled as much as 1.3 billion gallons out of the ground annually to transport coal slurry 273 miles by pipeline to the Mohave Generating Station in Laughlin, Nevada. Until the mines were shut down in 2006, that process used 4,400 acre-feet of water per day simply as a transport mechanism, depleting the Navajo Aquifer of water that would otherwise have been available to the artesian springs around Navajo and Hopi villages. The net effect was that fewer and fewer spring-fed fields and gardens had sufficient irrigation water, even as drought diminished the rainfall and runoff available to dune and floodplain fields.

This depletion of life-giving springs deeply concerned Navajo and Hopi elders and activists, who successfully lobbied for limiting the use of Navajo

Aquifer water by Peabody. "When Hopi elders say *Patuwaqats*—water is life—it is not a cliché, but a fact of life," explained Leonard Selestewa, a Moenkopi farmer who formerly served as president of the nonprofit Black Mesa Trust. "Water is not a commodity to be bought, sold, or wasted. . . . Water is sacred, especially in the Black Mesa region, where water is key to our survival."

By 2002, the diminished capacity for the Navajo and Hopi to farm with the use of spring water was becoming painfully evident. That year, the Hopi Pu'tavi Project collaborated with Iowa State University in a survey of some seventy-seven Hopi and Tewa farmers on the Hopi Indian reservation in Arizona. Of the sixty-four farmers still active on the three mesas and in Moenkopi, nearly 60 percent admitted that they no longer grew enough corn to meet their family needs. Although the Upper and Lower Moenkopi had a relatively high percentage of farmers—83 percent—who grew enough corn for family nutrition and ceremonial obligations, in villages such as Shungopovi and Tewa less than one-fourth of the farmers were corn self-sufficient anymore. Among all remaining Hopi and Tewa farmers, only 34 percent still grew peaches, 22 percent apricots, 21 percent apples, 12 percent pears, and 7 percent grapes. That entire set of crops had formerly been grown by nearly every Hopi and Tewa family. When the Natwani Coalition of Hopi farmers, educators, and activists did another survey of household food sources in 2005, over 80 percent of the food eaten in homes was purchased in off-reservation cities.

In 2005—at the height of the drought that began in 1996—Shawn Kelley and I were volunteering for the Natwani Coalition, and we showed several Hopi farmers the crop survey that Whiting had done some seventy years before. They eagerly looked over the survey and volunteered information on which of the cultivated varieties and crop species named in Hopi were still grown by them and their clansmen. Of some sixty-three named varieties or species of fruits, vegetables, grains, and gourds traditionally grown by the Hopi by 1925, only thirty were still easily found in Hopi gardens, fields, and orchards. Other introduced crops that never had formal Hopi names—from turnips to cabbage to peanuts—had also disappeared. In short, just 47 percent of the food-producing varieties grown by the Hopi farmers during their era of food self-sufficiency were still being grown and shared among them. Between Vavilov's visit and the present day, the Hopi have lost not only much of the wellspring of their life, but also more than half of their traditional agricultural biodiversity and most of their capacity for local food security.

What is striking about the list of Hopi varieties that have been lost or greatly diminished since Vavilov's time is that they represent some of the more drought-sensitive crops that require spring irrigation in terraces rather than dry-farming to be productive. Runner beans, Aztec beans, turnips, cabbages, tomatillos, cilantro, cotton, tobacco, and safflower are among those species or varieties now absent from the Hopi, and they were formerly grown with irrigation below springs. The spring-fed terraces that formerly adorned the slopes below every mesa village are largely in disrepair, or at least nowhere near their former grandeur. The springs at Moenkopi were found to be contaminated with benzene from a Tuba City auto garage; most others remain relatively uncontaminated but have suffered from severely diminished flows.

Since the current drought began in 1996, fewer and fewer farmers have been planting the entire acreage of their dry-farmed fields below the mesa-top villages. While the drought itself is not as severe as the ones in the 1930s and 1950s, global climate change has generated a hotter, longer growing season in the Painted Desert, placing most crops at risk due to moisture deficits. When I passed from Second Mesa to Moenkopi in July of 2007, nearly half the fields I saw were under planted or had already lost a significant proportion of their seedlings to wildlife predation. The remaining plants, especially those of corn, showed the classic signs of drought stress: delayed, stunted growth; an upright angle to leaves that made them look almost spiny; and discolored, "burned" stalks and leaves. The fields looked as dry as when Vavilov visited them but ultimately yielded far less food.

Environmental changes are not the only pressures on the Hopi and Navajo food system, however. While interviews suggest that almost 12 percent of Hopi farmers remain worried that extended drought could disrupt the continuation of farming for their households, 22 percent were concerned about the lack of interest in farming among the younger generation. I have wondered if this sociological shift—one that has affected the future of farming in nearly all developing countries—was even fathomable to Vavilov. With so many young adults opting for salaried work in nearby cities or for government jobs, who is left to maintain their families' fields? Today, America educates its rural youth to aspire to be anything but a farmer.

For such reasons, the Natwani Coalition formed in 2004 to bring youth and young adults back into activities that help to renew Hopi and Tewa farming traditions around the mesas. At the organization's first Hopi Food Summit, I was heartened to see hundreds of Hopi—literally, from eight to eighty years of age—intent on revitalizing the local Hopi food system. They have since renovated the ancient terrace gardens and orchards at Wepo Springs, built greenhouses to propagate fruit trees and salad greens, and grafted hundreds of scion-wood prunings from old Hopi fruit trees onto hardy rootstock for later transplanting into orchards. Similarly, a diabetes prevention project has involved hundreds of Navajo and Hopi children in tending a "healing garden" full of healthy foods and medicinal plants on the grounds of the Indian Health Service Clinic in Tuba City. In the fall of 2006, hundreds of Navajo and Hopi attended a Native Foods Symposium held at Moenave, eight miles from Tuba City, where native and Hispanic heirloom crops such as peaches and amaranths still flourish.

In his 1930 lecture in Arizona, Vavilov admonished Americans for leaving so much of their country's agricultural biodiversity "undiscovered," and ultimately unprotected and vulnerable, in the sense that it was largely unrecognized and undervalued. Ironically, his brief visit to the American Southwest allowed him to visit the place in the United States where the greatest agrobiodiversity north of Mexico had been documented, and where that diversity clearly contributed to food security. Yet, because its value remained underrated, that crop diversity and the freshwater springs that supported it fell into neglect. Government policies have encouraged the Navajo and Hopi to become so dependent on cheap, surplus food resources grown elsewhere that it has undermined their own cultural motivations to remain food self-sufficient.

Several studies have now verified that the imported foods that have replaced the native crops in their contemporary diet are poorer in protein, minerals, and dietary fiber but flush in fats and sugars. Over the decades that native crops have been replaced in the diet by simple carbohydrates, more and more Hopi and Navajo have been afflicted with adult-onset diabetes, obesity, cancer, and heart disease. Belatedly, they are rediscovering that the best cures for those maladies may be the very crops that once grew outside their own back doors.

That pattern of accelerating crop loss since World War II—and of recent attempts to recover former elements of local food diversity—is not at all unique to the Hopi, Navajo, or, for that matter, to the indigenous peoples of North America as a whole. Over the entire continent, more than a thousand unique seeds, breeds, and wild food populations have become threatened or endangered over the last century, dramatically reducing the diversity and resilience embedded in American food systems. Yet, as the Renewing America's Food Traditions initiative has recently documented, all is not lost; many successful efforts to bring some of the historic foods back onto tables in homes and restaurants across the continent are already under way. It is once again a time when Americans are discovering the richness of their own continent, just as Vavilov once admonished us to do.

CHAPTER TEN

Logged Forests and Lost Seeds: The Sierra Madre

"The mountainous areas of Central and South America—the Cordilleras—" announced Vavilov in 1925 at the age of thirty-eight, "are of exceptional interest to us as globally significant centers of origin fundamental to understanding the evolution of a number of important plants." It seems that those regions were of personal interest to him, as well, for in a letter written that same year to Sergei Bukasov, he expressed a bit of envy that his colleague was able to explore Latin America before he himself had a chance to do the same: "For me, Mexico is a country of great interest: The history of its agricultural crops, the composition of its cultivated plants, the complexes of maize, tobacco, solanaceous plants [tomato, potato, and other nightshades], beans and gourds are all new to me. What do they really represent? What do *you* find in the markets of the towns? Do you take photographs? . . . Do you keep a diary?"

During his 1930 visit to the Americas, Vavilov made good on his pledge to follow Bukasov's earlier route down into Central America, to immerse himself in some of the most remote and rugged regions of Mexico and Guatemala. After a month in Arizona and Southern California, he left the United States on October 30, heading for Mexico City, edging along the western slopes of the Sierra Madre for much of the way. Along that route, he saw dozens of species of pines, oaks, and agaves growing atop the volcanic cliffs, or *cumbres*, of the sierras. He tasted dozens of varieties of maize, eaten right off the cob as they were maturing in the fields, or shucked and soaked in lime to make hominy-like *nixtamal* for tamales, posoles, or tortillas gorditas.

Just as he had done on other continents, Vavilov offered up from his travels to the Americas a litany of names of endemic crops hardly known to science, let alone used in other corners of the world. But then he took a Nureyev-like leap:

> *Mountain regions* are the primary centers of origin for the crops I have enumerated. It is there, as our research has demonstrated, that an exceptional wealth of crop varieties lies hidden. . . . [We still have] *the opportunity to discover a great new "America" there*. . . . I have given myself the task of trying to discover the areas with the greatest accumulation of diversity in crop varieties native to Central and South America. The journey along the Cordilleras has offered me a chance to fulfill this task, and now I am able to pinpoint with great accuracy where this diversity lies. . . . [emphasis added]

It was in Mesoamerica that Vavilov identified one stretch of the cordilleras—the Sierra Madre Occidental—as America's mother lode of food biodiversity. As the "Mother Mountains" running parallel to Mexico's west coast, this particular cordillera forms most of the continental divide from the U.S.-Mexico border nearly all the way south to Guatemala. The volcanic ridges of the Sierra Madre Occidental form the edges of a half dozen canyons nearly as deep and just as long as the Grand Canyon of the southwestern United States; from the patches of tropical palms on the bottom floor of those gorges, you can see pines high above you on the rocky rims.

Along those steep gradients Vavilov fully elaborated a hypothesis that he had first roughed out back in the Pamirs of Central Asia: that the topographic heterogeneity found in mountainous regions serves to harbor such high levels of biological and cultural diversity that they foster the origin, evolution, and divergence of crop varieties. He recognized that the same geographic factors that generate new wild species in the cordilleras—impenetrable physical barriers, steep climatic gradients, and bizarre juxtapositions of soils—also promote a diversity of varieties within the same crop species. This is especially true when the crop has long been nurtured by indigenous communities such as those nestled deep within the Sierra Madre.

We now recognize that the mountains of western Mexico are part of a larger mosaic shaped from two global centers of diversity, the more tropical Mesoamerican lowlands such as the palm-lined canyon bottoms, and the more temperate and subtropical Madrean pine-oak woodlands found up on the canyon rims. The former harbors an astonishing 17,000 plant species—18 percent of them endemic—plus 440 mammals and 1,120 birds. The pine-oak woodlands above harbor 5,300 species of plants—75 percent of them endemic—as well as 330 mammals and 525 birds. In Vavilov's time, there were few inroads into the Sierra Madre to give biologists a chance to encounter so many species, let alone name and map their distributions. The sheer ruggedness of the sierras and the poor transportation infrastructure of his era hindered his efforts to delve deeply into that montane region, but he skirted its edges in a number of places, using trains, cars, and pack mules to the best of his ability to gain an overview of the habitat heterogeneity there.

Life in any mountain landscape is much more complex than initially meets the eye. Once the seeds of a plant are dispersed into a remote canyon or watershed, the rugged terrain surrounding the patch of seedlings isolates them from others of their own species. The peculiar microclimate of the new habitat, with its attendant bugs, birds, and microbes, places new selection pressures on the plant population, and it sooner or later diverges from its source population. However, periodic floods, windstorms, fires, or landslides sometimes bring other seeds or pollen in from other areas, allowing some hybridization to occur with novel sources of genes. This recurrent swing between isolation and hybridization engenders a process called *reticulate evolution* and makes such canyons ideal nursery grounds for speciation and diversification.

Such phenomena are of interest not merely to biogeographers, but, at least metaphorically, to cultural geographers as well. Linguists such as Joanna Nichols have recently articulated another reason mountainous regions today tend to be richer than other regions in species, crop varieties, and languages. Compared to extensive plains and coastal valleys, they are not as immediately susceptible to rapid colonization by new settlers, imperialistic domination, ethnic homogenization, and agronomic monoculture. In other words, Vavilov would undoubtedly find seeds in the Sierra Madre that had already been lost from the more accessible valleys and plains below them.

While passing across the bridges spanning the gorgeous subtropical barrancas below, Vavilov stared wide-eyed at the chaos of wild palms, climbing bean vines, and strangler figs, as well as corn's closest relative, teosinte. He visited isolated *rancheria* farmsteads and multicultural vegetable markets. He also walked along the ancient trails between the milpas of indigenous cultivators, who sowed their seeds on the fertile floodplains that edged the raging rivers spilling out onto Mexico's Pacific coast. They offered him swigs of their corn beer or tequila-like mescal, pouring him cup after cup of the ancient brews, which they stored in bottle gourds and ollas of crudely fired clay.

At his southernmost destinations in Mexico and Guatemala, Vavilov spent considerable time perusing prehistoric Mayan ruins, contemplating the grandeur of some of the most ancient agricultural civiliza-

Vavilov in the field in Mexico, probably 1931.

tions in the New World. But he also conversed with living descendants of those early agricultural innovators, bantering and bartering with the Zapotec traders of seeds, the cooks of toasted crickets, and the vendors of the myriad vegetable varieties stationed in the sprawling open-air market of Mitla, Oaxaca.

The photos of the Mitla region are among Vavilov's best. They feature more than just the wild and cultivated plants of Mexico; they document indigenous farmers tilling their fields with what he termed "Egyptian plows"; boys using pots to irrigate their gardens; elders tending native bees that serve as pollinators; women storing their harvests; and whole families going to market to hawk their wares. Perhaps nowhere else did he focus his lens so intensely on the simple face of humanity expressed in a center of diversity.

Near Mexico City, a farmer wearing a multicolored poncho and a huge straw sombrero led him through his milpa of maize crowded with teosinte. Looking less like modern-day maize and more like a tall, spindly grass with tiny wedge-shaped grains, wild teosinte has remained present in the fields of fully domesticated corn for some eight thousand years, ever since maize and teosinte first diverged on their evolutionary paths. Vavilov was just vain enough to have his colleagues take photographs of him hugging mature plants of maize and teosinte, which his colleague Sergei Bukasov had documented in 1925 as spontaneously hybridizing with one another. Bukasov and Vavilov correctly inferred that Mexican teosinte was the closest wild living relative to corn in all its various forms. Subsequent studies by molecular biologists, population geneticists, and ecologists have obliterated the notion that the wild ancestor of maize is now extinct, placing certain fully wild teosintes at the very base of corn's family tree. By comparing teosinte to the diverse strains of corn in the markets of the sierra, Vavilov came to understand viscerally the simple aphorism my colleague John Tuxill later coined: "Crop biodiversity is the biodiversity that people made."

Satisfied that he had at last glimpsed the reputed mother of maize—and had peered into the depths of the Mother Mountains—Vavilov returned to the United States by traveling up the eastern foothills of cordillera. His return trip northward took him across the Mesoamerican highlands north of Mexico City commonly referred to as the *altiplano;* its ridges were lined with terraces of sword-leaved agaves and giant prickly pear cacti. He lingered for a while in Chihuahua—perhaps in the early weeks of 1931—where he collected a nonflowering branch of the desert shrub known as guayule, so familiar to me from years of wandering through the American Southwest, inserted it into his plant press to dry, and then crossed into Texas at El Paso.

Some seventy-five years later, Sergey Alexanian took me through the VIR herbarium to show me a few dried specimens that Vavilov had pressed in Chihuahua. Those scrappy samples of desert plants triggered much more of an emotional response in me than they might in a casual observer. Guayule (*Parthenium argentatum*) was the first specimen he showed me and it was one of the plants so familiar to me. Vavilov's dried specimen seemed to exude a peculiar elegance. That seemed odd to me at first, because in the wild, the plant is a rather unkempt shrub with chaotic clusters of woolly leaves. Yet this particular specimen had been meticulously prepared by one who had mastered the art of mounting botanical specimens, and whose own flamboyant signature on the label—*N. I. Vavilov*—gave it an added flare.

The very fact that this specimen was collected, much less transported out of Mexico—and then endured the siege of Saint Petersburg through World War II—filled me with wonder. In 1931, several months after his first visit to Mexico, Vavilov had returned to its southern border and requested permission to travel up through Yucatán and on to the northern desert states, where he wished to collect guayule seeds. He was caught off guard by the response: For the third time in his career, he was arrested as soon as he crossed the border into an another country. U.S. diplomats and corporate interests in the Mexican capital had been monitoring his movements and had tried to prevent him from reaching the guayule rubber patches in the Chihuahuan Desert.

Why did Vavilov's interest in studying and collecting seeds of guayule trigger such apparent xenophobia? Vavilov knew that for more than a quarter century, Mexico had been exporting guayule to the United States, so that tires could be made there with an alternative to the hevea rubber plants of the tropics, which had been suffering from a blight. In fact, the Continental-Mexican Rubber Company had harvested and exported so much guayule rubber from northern Mexico that by 1912 its operation was both socially and environmentally unsustainable. As guayule historians have conceded, Continental's production ended due to "depletion of the native stands and civil unrest."

A year before his own venture into Mexico, Vavilov had seen the experimentally cultivated plots of guayule in California managed by the Intercontinental Rubber Company (IRC), an enterprise financed by the Rockefellers and Baruchs. Since guayule did not grow naturally in California, plantings there had been started from seed collected from the wilds of northern Mexico. But when the Great Depression began with the stock market crash of October 1929, IRC's investors were forced to shut down their marginally successful rubber production experiment so that they could reallocate their remaining wealth. Nevertheless, they were not willing to let the Soviet Union beat them to the punch in developing an alternative source of rubber. In the view of those Americans, Vavilov had to be stopped from taking seed back to Russia.

So, for the first time since his visit to Spain in 1927, Vavilov was being treated as an international criminal, a potential bio-pirate. It did not take him long, though, to deduce that the accusations against him were not really coming from Mexican officials; the shareholders of the U.S.-based Intercontinental Rubber Company were the more likely source of discontent:

> Later it became apparent that these difficulties were provoked [by] the Intercontinental Rubber Company of the United States, which was irritated by their knowledge of my errand in 1931 on behalf of the "Caoutchouconos" that underwrote my special expedition to Mexico to collect guayule. . . . [The American company] had started a campaign in the Mexican press about "the plundering of national treasures by the Bolsheviks."

The Caoutchouconos that Vavilov referred to ran an amalgam of Soviet rubber enterprises, which, like the American rubber companies of its era, were desperately attempting to find other sources of latex in the event that Germany or Japan cut off their access to the remaining hevea rubber plantations in the tropics. By the time Vavilov was arrested, IRC investors were already lobbying Congress to purchase their assets in guayule for some $2 million, arguing that securing an alternative source of rubber was of strategic interest to the United States. When those investors learned that the Soviets also had an interest in guayule research and development, they undoubtedly feared that Vavilov's well-trained team of economic botanists could outcompete their own.

In the end, it appears that Vavilov convinced Mexican officials that if U.S. corporations had removed enough seed from guayule plants in the Chihuahuan Desert to plant thousands of acres in California, the cat was already out of the bag. He could not be a pirate if he had explicitly asked their permission to take plants with him for long-term research, while other countries were already economically developing the same plant. Mexico temporarily backed off on restricting further collection of guayule and apparently granted Vavilov permission to gather at least herbarium specimens of the shrub. However, when World War II erupted a few years later, Americans once again put pressure on Mexico to be more restrictive with its natural sources of rubber. In 1942—while Vavilov sat locked in jail, starving to death—Mexico acquiesced to the U.S. government's demand to be the sole purchaser of all Mexican production of guayule.

Ironically, within months of the end of World War II, the formerly ailing American rubber companies were "feeling their oats." They demanded that the U.S. government destroy all of its guayule fields on both sides of the border, abandon all of its crop improvement and rubber extraction research on the shrub, and provide them with the political and military support to regain access to hevea rubber plantations in any tropical country where they wished to work. They wanted nothing less than exclusive control of the rubber industry through the tropical plantations of hevea they had appropriated. Within months, all guayule plantations in the U.S.-Mexico borderlands were torched by the U.S. government, and all but one barrel of the guayule seeds improved by two decades of plant breeding were destroyed. Thirty years later, when U.S. government officials decided that it had been a mistake to suspend its entire guayule research program under pressure from the rubber companies, the man who had salvaged that barrel of seeds and guarded it for three decades offered to sell the improved guayule seed stock back to the government for $1 million.

But back in 1931, as news of Vavilov's detainment spread, Mexican officials were showered with telegrams and phone calls demanding his release. Mexican scientists offered both apologies and assistance to get Vavilov out of house arrest. No doubt Vavilov thought back to the prior winter, when he had first found guayule shrubs in the field.

By that time in his life, Vavilov was well aware that choosing which plants to harvest and which seeds to save was an inevitably moral act inseparable from the political and cultural influences of his time. He saw himself not as a plunderer of botanical treasures but as a conserver of future possibilities for humankind. Nevertheless, he knew full well that the act of collecting plants from one country for potential use in another was never ethically neutral. Whether he is ultimately seen as a pragmatic conservationist, a bio-pirate, or a botanical carpetbagger from the north intent on acquiring the culinary treasures of the south will depend on who is making that assessment and under what political context.

What ultimately motivated Vavilov to visit Mexico was not his interest in rubber, however important that was to his society in the short term. His ultimate motivation was evident in his lifelong quest to learn where various foods came from, both geographically and genetically. Perhaps that quest is why he was so enchanted by finding corn and teosinte growing in the same milpa, and why I myself metaphorically followed in his footsteps. Perhaps that is also why I chose to retrace Vavilov's steps into Chihuahua, Mexico, and then turn westward, heading into the Guadalupe y Calvo municipality of the Sierra Madre. That is where I had first seen maize and teosinte growing together in the same field some two decades earlier, and it was an area where considerable logging had gone on in the meantime.

The Sierra Madre remains a linguistically diverse landscape, where both the Tepehuan (*Odami*) and Tarahumara (*Rarámuri*) still farm over ten thousand hectares of traditional crops, and where the endangered *Tubaris* language hovers on the brink of extinction. It is also one of the more diverse corn-growing areas of the cordilleras, for within a two-thousand-meter-elevation gradient, the Tarahumara and Tepehuan grow sixteen of the twenty-five races of maize known in Mexico. The vernacular names of those maize races make my mouth water just by saying them: *chapalote, reventador, dulce, conico, dulcillo de noroeste, elotes occidentales, conico norteño, tabloncillo, vandeño, chalqueño, cristalina de Chihuahua, blando de Sonora, onaveño, Pima-Papago, harinoso de ocho*, and *pueblo*. Each has a different taste, a distinctive set of uses. According to an incompletely sampled archaeological record, at least ten of those races of maize have been consumed in the northern Sierra Madre since prehistoric times.

In 1988—roughly fifty-six years after Vavilov's last visit to Latin America—I went into the sierras to see if the cross-pollination of maize and the northernmost race of teosinte was still occurring in Nabogame, Chihuahua. Almost two decades after that, in the spring of 2007, I had a hankering to visit other Tepehuan and Tarahumara rancherias of that region to see if their fields and granaries were still as diverse as I remembered them to be. While their fields remain nestled on steep volcanic slopes, and most of the Tarahumara as well as Tepehuan women still wear their bright, multicolored *trajes* as in centuries past, there have been changes wrought in *serrano* (mountain) culture and agriculture over the last century. You can find Tarahumara women in traditional dress, sitting on the ground shucking corn while listening to an American football game on a solar-powered radio.

Although Vavilov himself did not reach into the sierras and barrancas around Guadalupe y Calvo, a contemporary of his—the Norwegian-born explorer Carl Lumholtz—directed a major expedition for the American Geographic Society that passed through Nabogame and the Barranca Sinforosa in 1892, about the time Vavilov was born. In his 1902 book, *Unknown Mexico*, Lumholtz recounted his discovery of Tepehuan farmers intentionally mixing their maize with teosinte as a means of reinvigorating their corn seed. They call teosinte, the wild grassy relative of maize, *konkoñi usidi*—"wild turkey's stalks"—as if the turkeys planted its little seeds just as humans plant the large kernels of corn.

When my friend Garrison Wilkes went to Nabogame sixty years later, he confirmed that teosinte was still growing and cross-pollinating with maize on the edges of cornfields and in woody thickets along the streams just below them. I first journeyed into Nabogame in 1988 as part of a team from Native Seeds/SEARCH, a nonprofit agricultural organization that works on both sides of the U.S.-Mexico border. When I accompanied cofounder Barney Burns into Nabogame on mules, we had detailed directions from Garrison to help us find the teosinte there and names of Tepehuan families, as well. By that time, Nabogame teosinte was world renowned among corn geneticists and archaeologists studying the origins of agriculture, but the Tepehuan farmers we met were amazed that someone had once again come into their community to confirm that that wild grass still intermixed with the maize in their fields and among the trees along their streams.

Yes, they said quietly, tipping the brims of their straw cowboy hats, they remembered a gigantic American who came and spent some time in their maize fields in the mid-1960s. Was he married, they wondered about Wilkes. Yes, other Tepehuan farmers from as much as fifty kilometers away in "distant" barrancas sometimes paid them to grow their maize seed in the fields near where teosinte regularly germinates and grows. The "injection" or hybridization of teosinte made their maize kernels more flinty—good for grinding into pinole—and ensured higher yields for several more years. Yes, they said, they were pretty much growing the same criollo races of maize they had always grown. No, they shook their heads—a bit tired by now of such questions—they did not buy seed for planting from afar. Their teosinte-enriched corn seed seemed to do the trick.

While in Nabogame, Barney Burns and I made separate collections of seeds from about a dozen different corn plants and a dozen different teosinte plants growing in the same field. A few months after I sent them off to John

Doebley—a young geneticist who had already done as much as anyone to solve the mystery of the origins of corn—he sent me back a note with results that at first seemed to be full of surprises. Yes, he confirmed, there was genetic introgression between corn and the teosinte but only in one direction—teosinte pollen enriched corn but not the other way around. In other words, corn pollen was not available during the time when teosinte plants required it for fertilization and seed set. Though the flow of new traits from teosinte into maize was really a trickle, it was probably enough to generate some "hybrid vigor" that could potentially increase the yield of corn the following year, wherever it was grown.

In short, Barney, John, and I had reconfirmed then with more accurate tools what Lumholtz and Wilkes has suggested earlier—a continuing flow of genes from corn's closest relative in the Sierra Madre could explain part of the diversity found among the races of maize grown by the Tepehuan and their neighbors the Tarahumara, with whom they traded seeds.

It was now 2007, and eighteen years had passed since John and I published our field report in the *Maize Genetics Cooperative Newsletter*. I wanted to see just what had happened to the diversity of maize and the intensity of its use in the stretch of the Sierra Madre centered on Guadalupe y Calvo. In the intervening years, major changes had taken place in the region, but even more dramatic changes had taken place in the maize itself. When John Doebley had sought to confirm the gene flow between Tepehuan maize and teosinte in his lab in 1989, he had used isozymes to do so, not direct analysis of their DNA. But by the time I had returned to Guadalupe y Calvo, DNA analysis was commonplace in the research institutes of the United States and Mexico, and products made from transgenic or "GMO" corn could be found in nearly every grocery store in both countries, even though their sale was illegal in Mexico. I wanted to know if transgenic maize had already reached back into the milpas of the Sierra Madre.

In April of 2007, my wife, Laurie, and I found ourselves bouncing like ping-pong balls down the dirt roads of Chihuahua. We were in a van loaded with Mexican and American conservationists and Tepehuan and

Tarahumara human rights activists. Over the following two weeks, we rose as high as 2,800 meters in elevation when passing through the foothills of the Sierra Mohinora—the highest point in the Mother Mountains—and then nose-dived toward the tropical canyon bottoms of the barrancas, some of them lying at less than 700 meters in elevation. Where we could travel in one day would have taken Vavilov or Lumholtz five to six days in the 1930s.

We left pavement somewhere around the mill and mining town of Las Yerbitas, and for the next eighty kilometers, we climbed up pine-covered ridges and down into barrancas until we reached a cluster of 180 Tarahumara rancherias known as Choreachi, or Pino Gordo. It is a gorgeous valley of pasturelands studded with log cabins and granaries, with long volcanic ridges on either side. There, some of the last remaining old-growth pine stands hung on tenaciously to bluffs of pale volcanic ash, sheltering a rich understory of oaks, madroños, manzanitas, and wildflowers.

Below those ridges, the Tarahumara had freshly plowed their milpa cornfields in the bottomlands, *mawechi* bean patches on the lower slopes, and pasture grasses in the apple orchards that were tucked into canyons or drainages descending the ridges. The apple trees were already in full bloom, and the corn from last year's harvest was fermenting in large clay pots in the log cabins of each rancheria. For the next four days, whenever we approached their homes, the short-statured Rarámuri would offer us *jicara* cups full of the fermented corn beer known as *batari* or *tesguino* as their way of celebrating Easter. We would indeed learn much about the importance of the corn in their community and the factors leading to its erosion.

Earlier, I had the chance to discuss both the positive and negative changes in the region with Randy Gingrich, founder of the Sierra Madre Alliance, Barney Burns, and another Native Seeds/SEARCH cofounder, Mahina Drees. Randy, Mahina, Barney, and I had been among a dozen or so activists who took on the World Bank in 1985 when it proposed large-scale logging and paved roadways reaching far back into the Sierra Tarahumara, the branch of the Sierra Madre Occidental around Guadalupe y Calvo.

Although the World Bank dropped out of the region once our studies revealed the potential cultural and ecological impacts of its proposal, we could hardly claim any victory, since private interests had since developed much the same infrastructure of roads, saw mills, and airstrips. Those roads and airstrips were accelerating the rate of negative changes in the region, as we would soon see.

Since then, the northern Sierra Madre Occidental has been recognized as a mega center of plant diversity, ensuring that the Sierra Tarahumara is regarded by the World Conservation Union (IUCN) as one of the regions on earth most diverse in wild species, native crop varieties, and cultural traditions of sustainable use. This international recognition, we hoped, would slow down plans to log its ancient forests. Perhaps most important, Barney and Mahina led Native Seeds/SEARCH staff in a comprehensive identification and collection of native races of maize and other crops to safeguard them for future generations. During the 1980s, Barney and Mahina gathered some 175 samples of twelve races of maize from the Tarahumara. Many of these have since been returned to Tarahumara families, who lost their own seed reserves during a fierce drought in the 1990s.

The Tarahumara are today still a "people of corn." As Chihuahuan ethnohistorian Victor Martinez has simply and straightforwardly put it, "Maize is the backbone of the indigenous culture of the Sierra Tarahumara." Maize remains the keystone crop sown on three-quarters of all arable lands of Pino Gordo. Nearly a hectare of corn is still planted for every person in Choreachi, but during years of drought, floods, or plagues, the yield can be meager. In good years, the Tarahumara typically harvested 300 to 450 kilograms of edible maize per hectare, but in recent drought years, the harvests have declined to less than a fourth of the average yield. All told, a family of six Tarahumara and their livestock typically require about 4,700 kilos of maize to meet their nutritional needs. That's about 800 pounds per person, far less than the average hectare in Choreachi produces. Six months before the next maize harvest, most of the granaries we observed were empty of all but the seed corn saved for a May planting. Some of Choreachi's families said they did not even have seed corn left and would have to barter for some from their immediate neighbors or bring it in from farther away.

It appeared that for lack of arable land or lack of harvestable runoff from the forests above their fields, the farmers this particular year had insufficient quantities of corn seed for planting and maize for eating. By maize for eating, I do not merely mean fresh corn on the cob. Albino Mares Trías, a Tarahumara practitioner of traditional foodways, has demonstrated that nearly every part of the maize plant is eaten except the roots. The kernels of maize are more often than not dried rather than eaten in their fresh, green "milk" stage. Even when the entire corn on the cob is eaten, it is first cooked, then dried for five days before it is cooked again, then eaten with meat and tortillas as a dish called *chacal*. For other dishes, the shelled kernels are soaked, then boiled, toasted, and coarsely mashed with water and *quelite* greens or roasted mescal for *esquiate*; ground finely to make a coarse flour called masa harina; or toasted and ground to make pinole with flint corns and *atole* with softer flour corns. The corn husks are used to wrap masa, meat, and greens into tamales or the leaves of the native *makuchi* tobacco into cigarettes; and the dried corn tassels are boiled in milk or water to make a sweet drink. The cornstalks of *maiz blando* (but not other varieties) are sweet enough to be eaten like sugar cane.

When the maturing ear gets infested with corn smut, the smut's mushroom-like fruiting body is harvested along with the bloated corn kernels as a *huitlacoche*. The corn beer we imbibed on Easter Sunday was particularly nutritious and is leavened with a strain of beer yeast known only from the clay pots that the Tarahumara use to ferment *tesguino*. In short, corn is eaten, drunk, and smoked; like another essential, water, it is in every cell of the Tarahumara people who hosted us, coming into their bodies as a liquid, as a solid, and as a smoky vapor.

This cultural dependence on corn makes the recent shortages of traditional maize varieties even harder for them to bear. In Choreachi, we wistfully witnessed villagers making tesguino using bags of "instant masa" cornmeal of the Maseca brand.

Maseca is produced by a multinational corporation known as GRUMA. GRUMA Mexico—itself partially owned by Archer Daniels Midland (ADM)—together with ADM's U.S. branch co-owns the Azteca Milling plant in Edinburg, Texas. That mill allegedly provided the genetically contaminated flour from transgenic StarLink corn that caused considerable controversy in 2000. StarLink was not yet approved for human consumption when it somehow showed up in Kraft taco shells; it has since been replaced by other genetically modified corns in the marketplace.

The governments of both the United States and Mexico, and the corporations that dump U.S. corn into Mexican markets, would like us to believe that such problems are fully behind us. However, there have been unconfirmed reports since 2000 that transgenic or genetically modified (GM) corn also has made its way into Mexico illegally, perhaps through the various brands of corn chips.

Well before the reports on GM corn chips were released, GRUMA announced that its Maseca brand tortillas would be free of genetically modified organisms (GMOs), in accordance with Mexican law. But Greenpeace México claims that Maseca continued to use transgenic maize imported from the United States in its corn products sold in Mexico well into 2006. Maseca has also been criticized by Greenpeace for using Mexican government subsidies to dump cheap U.S. corn into the Mexican marketplace, even though its billboards in Chihuahua claim that it uses "Maiz de esta tierra," or "Corn of this country." In bringing a consumer fraud lawsuit against Maseca in June of 2006, Areli Carreón, the consumer advocate for Greenpeace México, argued that because seven out of every ten tortillas sold in Mexico come from GRUMA's Maseca mills in the United States and Mexico where batches of corn from various sources may be mixed up, GRUMA can still not guarantee that the maize used to make them is GMO free:

> While Maseca pushes a big advertisement campaign [claiming it no longer uses transgenic corn in the United States] among Mexican-Americans, back home Maseca is using [transgenic or GMO] corn to feed their families. No company

should be allowed to lie, exaggerate or deceive the public through their advertisements, especially if this company produces the basic food for Mexicans.

Unfortunately, GMOs have not only contaminated processed corn foods coming into the Sierra Madre, but there is growing speculation that they may have also contaminated the indigenous fields of diverse maize varieties there, as well. It is not known how much transgenic corn has gotten into traditional maize plantings in the Sierra Tarahumara, but a 2003 field sampling in fifty-three indigenous communities suggested that contamination of native corn by StarLink had already occurred in six states, including Chihuahua. Victor Martinez has reported that 33 percent of the samples taken from indigenous fields in the Sierra Tarahumara may be contaminated. While other reports of the 2003 survey do not specifically note the fraction of the Sierra Tarahumara samples contaminated by GMOs—that is, by transgenic maize releases from biotech firms—they do document physically deformed plants, as well as biochemical evidence of GMOs, in the Tarahumara homelands of the Sierra Madre. The sampling and testing techniques used have been criticized by some scientists, but concern remains, since so many Chihuahuan farmers purchase seeds in the U.S. when they make visits to bordertowns.

Pedro Turuseachi, a Tarahumara spokesperson with Chihuahua's Consultoria Técnica Comunitaria, had this to say about why the possible presence of transgenic corn is so threatening to his people: "Our seeds—of our own maize varieties—form the basis of any food sovereignty we have for our communities. Maize for us is much more than a food; it is part of what is sacred for us, part of our history, our currency, and our destiny."

In underscoring the importance of maize to every aspect of Tarahumara culture, Turuseachi argued that potential contamination of traditional maize varieties is not merely a technical "genetic" issue but a cultural and spiritual one, as well, since something as sacred as maize should not be desacralized by "impurities."

Maize for the Tarahumara and Tepehuan is many things, but there are also many maizes, each one of them fitted to a different microenvironment and use. Over the last fifteen years, the Native Seeds/SEARCH project called Treasures of the Sierra Madre has returned many kinds of native maize to farmers who lost them during droughts, while also helping Tarahumara families stabilize and restore many of the special microenvironments where those maizes were formerly grown. The Chihuahuan coordinator of the Treasures project, Juan-Daniel Villalobos, has assisted several Tarahumara communities and dozens of families with building stone-lined terraces, which help to retain the soil fertility and moisture required to grow traditional maize and beans. Where the terraces have been put in place, Juan-Daniel and his colleague Suzanne Nelson have reintroduced well over a dozen native maize and bean varieties that the Tarahumara farmers identified as once being prolific in their areas.

Despite such active efforts by the Tarahumara to maintain and regenerate their diverse maize legacy, some varieties are now considered to be endangered by all the recent introductions of particular hybrid maize cultivars. Even the notion that there might be one superior maize cultivar that will meet all community needs is considered to be a folly among the indigenous folk of the sierras; nevertheless it remains the pipe dream of some plant breeders.

Just three years after Barney Burns and I visited the Guadalupe y Calvo rancherias in 1988, the Chihuahuan state government began to distribute two "improved" (but non-GMO) cultivars of corn to the Tarahumara and Tepehuan of that region. The Vanta-1 and Vanra-1 cultivars were developed from crosses between a criollo maize from the sierras combined with others that were higher yielding (especially when fertilized) and produced larger grains and multiple ears per stalk. Over the last decade and a half, the large, white-seeded cultivars have been distributed by the state coordinator for the Tarahumara to dozens of highland communities.

The problem is that large, white, bland-tasting kernels are not very useful for the entire suite of foods that the Tarahumara and Tepehuan make from their diverse varieties of corn. They are virtually useless for making pinole, and the famous blue corn tortillas of the Tarahumara cannot be produced from them without adding food dyes. The blue anthocyanin pigments of highland maizes, as John Doebley once explained to me, are not merely for show. Blue and reddish purple pigments in highland maize seedlings absorb more heat early in the day and in the growing season, when cold temperatures may otherwise stunt the development of corn plants. Therefore, by distributing only white corn seed, the plant breeders are selecting *against* one of the most time-tried adaptations of maize varieties in the highlands of the Sierra Madre.

The more widespread varieties of traditional Tarahumara and Tepehuan maize remain viable and well cared for by many, Juan-Daniel Villalobos explained to me, but seed of the more place-specific varieties with special uses are growing increasingly rare. That may also be true of the Nabogame race of teosinte, which remains known from just three localities in the sierras; some have estimated that their total range is now less than fifty square kilometers.

Both traditional maize diversity and wild teosinte are unfortunately also vulnerable to two other pressures that have been mounting in the Sierra Madre for more than seven decades: logging and drug production. These two disruptive forces have been tearing the ecological and cultural fabric of the sierras.

Although the first European and mestizo miners settled in the northern Sierra Madre around 1708, almost all their logging was done for local uses, the building of homes and mine shafts and the burning of wood as fuel. After 1884, however, Mexican governmental policies fostered the building of railroads and company towns and the extraction of timber for extra-local uses. Around the time of Vavilov's visits to Mexico, U.S. investors began to back Chihuahuan mestizos to extract wood pulp and cellulose for U.S. markets. After World War II, logging companies made more and more inroads to remote parts of the sierras, in many cases taking lands away from indigenous communities so that they could be clear-cut. Today, as Randy Gingrich has written in the reports of the Sierra Madre Alliance (SMA), "Over 99 percent of the original old growth forests of the Sierra have been logged. Secondary forests lack the structure to sustain biocultural diversity—studies sponsored by SMA have indicated a significant loss of biocultural knowledge in secondary forest areas."

Where logging roads and airstrips were introduced in a region, other kinds of changes then filtered in—changes that erode the soils, the traditional ecological knowledge of indigenous farmers, the diversity of their crops, and the integrity of habitats surrounding them. One of those changes is the usurpment of arable land formerly dedicated to maize and beans by the sowing of opium poppies and marijuana. The narcos ride in to the region in their four-wheel-drive Ford Escorts or fly in with their Cessna Cubs and offer indigenous campesinos *a thousand times* more than what they can sell their corn for if they will grow drugs instead.

That's right, a thousand times the income from one hectare than what a poor farmer has ever made off his corn. According to Chihuahuan economist George Mayer the salable harvest from one hectare of Tarahumara or Tepehuan maize has seldom garnered the farmer more than five hundred pesos per year; marijuana from the same hectare will render as much as five hundred thousand pesos, if indeed that much cash ever stays with the farmer himself. Sadly, I had to ask myself, if you were in their sandals, what would you grow?

In essence, logging has become an economic "cover" for a much more lucrative business—drug production and trafficking—which now generates more income than all the other economic activities of the sierras combined. Road building and deforestation—and the cultural and ecological destruction that come in their wake—continue at a pace that Randy, Barney, and I could not have imagined when we thought we had "won" our case against the World Bank's promotion of logging in the Sierra Tarahumara.

While I was in the Guadalupe y Calvo reach of the Sierra Tarahumara in April 2007, the staff of SMA was anticipating a judge's decision in Chihuahua City on whether he would block plans to let non-Indians immediately log most of the remaining timber in the Pino Gordo *ejido* (collective), in what has been considered the last old-growth forest of any size in the Barranca Sinforosa, a major valley. Randy Gingrich was on pins and needles the days we traveled together, but the decision kept being postponed. His anxiety seemed justified, once I learned from him what was at stake.

"Maybe only 5 percent of the commercially suitable trees around Choreachi have ever been cut, but if the judge doesn't support our case for maintaining indigenous community control over these resources," Randy sighed, looking down at the ground, "90 percent of the old growth here will be gone in three years." The case remains in review and is hotly contested. The activists associated with SMA are not the only ones concerned about the impending threats of accelerated deforestation and road building for drug traffickers on the biocultural diversity of the Sierra Madre. Two eminent scholars of traditional ecological knowledge in the Sierra Tarahumara—Serge LaRochelle and Fikret Berkes—have recently expressed the same concern:

> Perhaps the major threat [to traditional uses] by the Rarámuri is the loss of control over the forest commons. Increased activity [of miners, loggers, narcos, and tourists] has impacted local ecological relations, and contributed to deforestation, soil erosion, and the loss of understory plants. Such changes are threatening the traditional ecological knowledge and the cultural integrity of the Rarámuri people.

There is also mounting evidence that the ecological and cultural changes that have occurred in the sierras since Vavilov's era are threatening the mother of corn in the Mother Mountains. At a 1995 meeting sponsored by the International Maize and Wheat Improvement Center at El Batán, Mexico, three prominent agricultural scientists all expressed their concerns about the future of Nabogame teosinte.

Garrison Wilkes considered Nabogame teosinte to be rare, with its historic distribution contracted. He suggested that the spread of roads into remote areas, the isolation of small teosinte populations, and the introduction of new cash crops (like marijuana and poppies) have cut teosinte's distribution in half since Vavilov's time. Two of Wilke's Mexican colleagues have made botanical pilgrimages like ours to Nabogame. In their independent assessment, Jesús Sanchez and José Ariel Ruiz determined that the teosinte there was rare and "threatened by deforestation." They also suggested that the current range for that race of teosinte may already be reduced to as little as thirty square kilometers. About the same time, geneticist Lesley Blancas reported that the majority of subspecies or races of teosinte in Mexico are now in danger of extinction.

Every once in a while, when I am back home in Arizona, I look at photos I made in 1988 when I first harvested a handful of teosinte kernels from the Tepehuan fields of the Sierra Madre. One photo is a close-up of the wedge-shaped grains that neatly fit together like pieces of a jigsaw puzzle. The light on the little cluster of grains made them shine, as if they were almost glowing against the black backdrop on which I had placed them. As I look at them now, they offer me a glimmer of hope surrounded by a world of darkness. No wonder Nikolay Vavilov wished to hug the corn and teosinte plants that he met in the fall of 1930 on the outskirts of Mexico City, a place, no doubt, that is now surrounded by paved roads and populated homes. Perhaps he wished to hang on to them for just a moment, with the hope that somehow they might stay with us forever.

The dynamics of natural hybridization between maize and teosinte are perhaps peculiar to Mexico and Guatemala, but genetic contamination of ancient cereal grains, vegetables, and fruits by transgenic cultivars is a new dynamic and one that is becoming increasingly commonplace. Farmers may temporarily enjoy higher yields when they adopt certain GMO crops, but more and more case studies indicate what they are losing, not just what they gain. Whether they are canola farmers in North America, sorghum farmers in Africa, or rice farmers in Asia, more food producers around the world now see that by uncritically adopting any transgenic crop that becomes available to them, they may lose control of the way their crops and certain weeds have positively interacted over many millennia.

Many prominent biologists now point out that the heady claims about how transgenic crops would put an end to hunger have already begun to wither under careful scrutiny. As conservation biologist David Ehrenfeld recently summarized the situation,

> Despite the enormous popular enthusiasm whipped up by the press and the financial markets, only a small proportion of the simplest possible genetic manipulations among the many that have been tried have worked at all. And many of these have turned out to be disappointing, dangerous, or both. . . . It has become increasingly apparent that DNA is only part of the story [shaping a particular crop's success, because] it is subject to other regulating and modifying influences in the cell, influences that we hardly understand. . . . In other words, the idea that patented transgenic organisms (and there are now many) are genetically stable and capable of performing consistently as desired for long periods of time and through many generations, is not biologically warranted.

Warranted or not, the idea that transgenic crops will feed humankind in the future has now been "seeded" in nearly every center of diversity that Vavilov once visited. Even if the transgenic grains, fruit, and vegetables have not physically arrived in the fields, orchards, and gardens of these mountainous regions, it is only a matter of time before their presence will be felt in the food system of each of these centers of diversity. Corn chips, for instance, are transported from country to country with little recognition of the ingredients within them. Likewise, transgenic seeds may be cleaned and bagged in the same mills where traditional seeds are cared for, and they can inadvertently contaminate the next batch of seeds that runs through a cleaner. The genies have been let out of the lamp. And so have the genes.

CHAPTER ELEVEN

Deep into the Tropical Forests of the Amazon

There could not have been a tougher time in Soviet history for Nikolay Vavilov to undertake an expedition to South and Central America than 1932. And it turned out to be the last time he was allowed overseas. Even before he took a leave in late 1931 from his duties in managing some twenty thousand researchers in 155 experiment farms and some 250 other research sites, the Soviet Union had begun its free fall into the worst famine since the Russian Revolution.

Although he undoubtedly saw some signs that the famine was coming before his departure for the American tropics in the summer of 1932, Vavilov was intoxicated by the enticement his colleague E. V. Vulf had given him to visit the rain forest: "Concealed in the bosom of the tropical floras are inexhaustible riches . . . unknown and valuable plant products." Despite many signs that he should stay put and steer the ship then known as the All-Union Institute of Applied Botany and New Crops, he ultimately decided to trust his wanderlust, as he had in the past.

That choice would ultimately cost him his life. Of course, he could not have fathomed at that time that the coming famine would ultimately result in 400 million of his countrymen going hungry, millions of them literally starving to death in the Ukraine and the adjacent regions of Don and Kuban. Bad weather triggered this famine, but political factors turned bad luck into a catastrophe of titanic proportions. As former Soviet official Victor Kravchenko has documented, Stalin's collectivization of his people's farmlands not only failed to increase crop yields, it directly resulted in mass starvation when farmers who saw their family's fields collectivized lost their incentive to work the long hours that competent farming inevitably requires. To make matters worse, Soviet bureaucrats sent troops to confiscate grain from the stubbornly independent farmers who had resisted collectivization by sowing fields on their own in remote locations. Because of such tensions with the government, two hundred thousand Kazakh peasants fled their country before the famine finally waned in 1934. Before Vavilov could return to the Soviet Union from the Amazonian rain forests and Andean highlands, between 2.5 and 4.8 million peasants had starved to death, victims of their own government's incompetence, inflexibility, and stupidity.

When Stalin realized the mess he had created, the megalomaniacal dictator began searching for a scapegoat, someone in the agricultural domain who could shoulder the blame. So much of Stalin's murderous wrath had been directed at Lenin's favorites like Kamenev, Zinoviev, and Trotsky that Vavilov presented an inviting target, for being earlier endowed by Lenin with one of the largest workforces for agricultural research the world had ever seen made him politically vulnerable. Rather than mobilizing his many workers to develop higher-yielding crops that could offset the lethargy of Soviet collectives, it appeared that Vavilov was off "sightseeing" in the Americas, leaving his management responsibilities to others.

Before he departed for the Americas for the last time, some of the first criticisms from Stalin's supporters had already reached Vavilov's ears. And while in the Americas, the secret police of the People's Commissariat of

Internal Affairs (the NKVD, precursor to the KGB) opened a file on Vavilov. Given that Vavilov almost certainly knew he was both under attack and neglecting vexing problems at home, it is not hard to detect an almost frantic desperation in his itinerary, for in less than eight months he dipped into fourteen countries: Cuba, Trinidad, El Salvador, Costa Rica, Honduras, Panama, Colombia, Suriname, Brazil, Venezuela, Peru, Bolivia, Argentina, Uruguay, and Chile. His schedule was so stiff that the only time he slept was while moving from collecting site to collecting site. His fellow geneticist Carlos Offerman curiously watched him fall asleep during a perilous plane ride through a lightning storm over the rain forests of the Brazil-Suriname border while other passengers were screaming their heads off. Surely he knew that he could not do justice to the hearths of early agriculture of so much territory in such a short period of time, but he kept repeating to his colleagues in the Americas a phrase that now seems tragically prophetic: "Time is short, time is short, and there is much to do. One must hurry."

Perhaps his desperation had been triggered by the increasing conflict he felt between achieving his vision of a world collection of plant resources that could be understood through an evolutionary framework, and his nagging duties as the head of an unwieldy agricultural research empire back home that was ultimately under the direction of a mad and bloodthirsty dictator. The year before—just as he was to leave the United States for his "dream trip" into the tropics of Mexico and Guatemala—he had received a telegram from a Soviet government official, insisting that he abandon his

Vavilov in the tropics in Venezuela, 1932–33.

fieldwork to attend a high-level political meeting. But he didn't change his plans. As he confided to American botanist Homer Shantz at the time, he believed he had been given a higher calling than to jump when bureaucrats asked him to jump. In paraphrasing Vavilov, Shantz reveals Vavilov's intellectual independence:

> This wire is from a man who is not above me. If I were a [functionary] in the Communist Party, I would have to obey. In that case, I could not use my own judgment. But I am employed by the Communists to work for the welfare of the people of the USSR, so I am still free to judge what is best. I will answer the wire [by saying that] it is more important for the future of the people of the USSR that I visit the centers of origin of cultivated plants in Central America than I attend any state dinner that can be arranged.

This commentary might indicate that Vavilov was oblivious to the realities of maintaining strong political support for his grand experiment in agricultural conservation and crop improvement. Nevertheless, the goal of improving and protecting the future of his comrades in the USSR was still foremost in his mind. Shantz never doubted that Vavilov was sincere in his belief that genetic diversity was the best means to ensure food security and nutritional well-being for his—indeed, for all—people. As Shantz later recalled, "We had been together some days and had many talks about the future of agriculture, of the desirability of knowing the natural resource of plant material to be drawn upon, and of the necessity of providing food for the then-starving people of Russia and other parts of the world. His whole desire was to improve the standard of nutrition for his people and those of the rest of the world."

Vavilov had a preternatural sense that there might be something altogether extraordinary about the plant diversity he would find in South America; and, since his death, hundreds of floristic, ecological, and biogeographic studies have verified that his intuitions were on track.

He intended to gain the inner reaches of the Amazon basin from both its Colombian headwaters and the sprawling river delta on the Atlantic coast, but the time he could afford and the tremendous logistical difficulties of roaming those remote areas severely limited his options. As days turned into weeks, he realized that he had only skirted the edges of the Amazon basin, despite meeting with dozens of Latin American as well as Russian diplomats who promised to help but had gotten him no closer to his ultimate goal. Finally, on that hair-raising flight over the rain forest that had his fellow passengers screaming while he slept like a baby, he was taken into the heart of the greatest expanse of rain forest on the planet. After being met by some Brazilian scientists who had recently established a field station to document the biodiversity of the Amazon, Vavilov got to glimpse a world that he had been hoping to see for years:

> We went deep within the tropical forest. It was necessary for us to be equipped with raincoats and umbrellas. However, the rainy weather we encountered quickly changed into a day bright with sunshine. When it rains in the forest, all falls silent, as if all life comes to a standstill. The rain would fall for a while, and then the sky would turn blue, allowing the sun to reign anew.

Vavilov's field notes shine with the excitement he felt at that moment:

> As the forest became full of life again, an incredible chirping of cicadas and a peculiar rustling and noise in the branches could be heard. A multitude of hummingbirds flew around us, as did different kinds of insects. Among them were amazingly large butterflies, colored red or pale blue like the mother-of-pearl; they could be seen flying and roosting here and there. It was difficult to reach them for collection because the forest floor was so sodden and boggy. Indeed, this was an authentic *rain forest* . . . the extraordinary wealth of plant life in the tropics is its true hallmark. Studying a modest patch of rain forest of just two hectares, the Brazilian botanists here have found 2,000 species of higher plants, roughly the same size of flora that we might find within an entire European country.

While many biologists of his time were so dazzled by the bounty of the rain forest itself that they paid little or no attention to its peoples, Vavilov appears to have kept his eye open for any signs of indigenous agriculture:

> The small communities of indigenous people who live in the rain forest appear to subsist largely on their crops of cassava, corn, rice and sugarcane. You can encounter indigenous people there who still practice a very ancient but elegant form of agriculture, sustaining their traditional lifeways by nourishing themselves on wild fruits, roots, fishes, birds, and monkeys. These—the original stewards of the rain forest—are very few in number; the majority of them have settled in to farm along the river's edge [where] two kinds of manioc have been brought into cultivation by them. One is so bitter that it needs to be processed in order to release its toxins before it can be consumed as food; the other is not so bitter, and it can be eaten fresh, without processing.

He made further comments on the cultivation of perennial cotton, pineapples, papayas, mangos, yellow sapodillas, black sapotes, *grumichana* "rose apples (probably *Szygium*)," and Brazil nuts. This was an altogether different structure of agriculture than he had seen elsewhere in the world, one dominated by perennial trees and vines in mixed plantings that virtually mimicked the structure of the rain forest itself. Decades later, ethnobiologist Darrell Posey asserted that the *apêtê* floodplain orchard-gardens on islands in the Amazon were so well integrated with their surrounding habitats that literally tens of thousands of hectares of food-producing cultural landscapes in the rain forest had gone unnoticed by tropical ecologists. Posey's colleague William Balee went so far as to argue that at least 11.8 percent of the "wildlands" of the Amazon basin was actually cultivated and managed by indigenous farmers over many generations, so much so that it is tempting to call the terra firma of the Amazon a managed garden. While other scholars have suggested that Posey and Balee overestimated the influence of Indians on the forest, it has become clear that indigenous peoples of the rain forest have shaped the spatial and temporal dynamics of their food-producing habitats on a scale far grander than most twentieth-century scientists could discern. It is to Vavilov's credit that during such a brief visit, he could see the rain forest and the managed trees within it.

My own glimpses of the Amazonian rain forest and its orchard-gardens have been just as fleeting as Vavilov's, but I had the good fortune to spend my brief forays into the tropics with pioneering conservation scientists Mark Plotkin and the late Al Gentry. As a field program consultant for the Amazon Conservation Team (ACT), I was able to visit the Ingano, an indigenous culture of the Amazonian headwaters. They, more than any other indigenous farmers, have informed my view of how rain forest agrobiodiversity has changed through time. I spent some time in the Ingano settlement of Yurayaco on the Rio Caquetá in 1999 and later came upon informative dietary studies of the same community accomplished by Camilo Correal, Tim Johns, and Harriet Kuhnlein.

It is doubtful that Vavilov saw much deforestation in the Amazon basin during the 1930s, but rapid clearing of the rain forests was painfully apparent nearly everywhere I went. My flight into the Amazon basin arrived in Florencia, Colombia, rather than in one of the coastal cities of Brazil, where Vavilov's sojourn into the still intact forests began. As my wife, Laurie, and I left Florencia in a four-wheel-drive vehicle that could maneuver the mud-rutted roads, we immediately witnessed firsthand the clearing of the forests to provide pasture for livestock, as peasants with chainsaws cut away nearly every tree in sight, leaving only a few towering kapok trees to provide minimal shade for some raggedy cows. The drone of chainsaws drowned out the birdsongs around us. Through such an incremental but insidious process of fragmenting one forest patch after another, the Amazon basin has lost a fifth of its cover since 1950. Notwithstanding those losses, the Colombian Amazon remains rated among the twelve most biodiverse regions in the world, and its many surviving tribes keep it linguistically diverse, as well.

As we passed the fragmented forests surrounding Florencia, we drove into the Andean piedmont, which is drained by two major tributaries, the Rio Caquetá and the Rio Putamayo, the latter forming the still contested borders between Colombia, Ecuador, and Peru. We were not far from the equator. The Ingano—or lowland Inga, as they are sometimes called—are one of seven tribes in this equatorial region that move freely back and forth between Peru and Colombia and, on occasion, downstream into Brazil. Faced with loggers and livestock producers on one side and cocaine growers and smugglers on the other, the Ingano sought help from the Amazon Conservation Team to ensure that stretches of their aboriginal lands would remain off limits to development. Some of their people have now resettled to the edges of a national nature park, which protects some 68,000 hectares of their homeland from mining, forest conversion to livestock pastures, and other forms of development incompatible with Ingano lifeways.

To reach the Ingano settlement of Yurayaco, Laurie and I took a large river taxi that speedboated us down the wide, meandering channels of the Rio Caquetá. After an hour or so, we spotted some dugout canoes and their Ingano boatmen, who had been waiting to take us inland. We snaked through some overgrown channels in sultry, backwater lagoons before the canoeists left us on somewhat solid ground, five kilometers away from the Yurayaco settlement. I began to see small, cultivated plots much like those that Vavilov had described downstream, on the Brazilian side of the watershed.

For the next two hours, we trudged in our mud-covered rubber boots through forests, gardens, and orchards managed by the Ingano, whose bamboo housing compounds were often scattered on the drier ridges above their plantings. A variety of tropical fruit trees, palms, and root crops seemed to trail out in every direction along the muddy pathways between their elevated huts. Cassava, corn, zapotes, yams, peach palms, papayas, plantains, and pineapples dominated their plantings, with a couple of dozen of other minor crops scattered among them. Chickens, pigs, and other small livestock foraged on the weeds at the orchard's edge and consumed the trimmings left over after various harvests.

When we arrived at Yurayaco itself, we were hosted by several Ingano women who had prepared their cassava and yams by roasting them as sliced vegetables over a wood fire or stewing them in cauldrons into savory soups with chicken and pork. Then came the cups of *chicha*—a fermented corn beverage much like a sweet beer or mead—and baskets containing all matter of tropical fruits. Despite heavy reliance on cassava, yams, rice, corn, and beans as starchy staples for several days running, I was impressed by the variety of fruits, fish, seeds, and greens that complemented those dietary mainstays.

Over the years following our visit to the Rio Caquetá with the Amazon Conservation Team, the ethnobiologist-physician Camilo Correal, of Colombia's Instituto de Etnobiología, has worked with Ingano collaborators to document some 166 species of foods in the local diet, from cultivated roots to palm fruits to insects, monkeys, and fish. Their survey confirmed—perhaps for the first time ever—the complexity of the original diet of Upper Amazonian peoples, for wild, managed, feral, and strictly cultivated species contributed to seasonally varying cuisines that relied on baking, roasting, grilling, boiling, fermenting, and detoxifying a vast array of over one hundred different foodstuffs. At the same time, many of the Ingano foods could not be pigeonholed into easy categories such as wild versus domesticated crops. For instance, the fruit trees seen along trails could have been anciently and intentionally planted or feral seedling offspring derived from fruit moved from orchards by monkeys and pigs. Their wild relatives could still be found on the forest edges, and the entire range of fruit sizes within each species ranged from small to large, tart to sweet, and dry to juicy. The subtle transition between garden and forest traditionally found around the settlements made me realize that the Ingano were managing many wild medicinal and food plants just as they were more obvious domesticated plants, such that both coexisted and intermingled in their gardens.

Camilo Correal explained this "gradient" between wild and human landscapes in terms that echo those of Vavilov, Posey, and Balee:

> For many years, most Westerners have considered rain forests as primary, unique and pristine ecosystems. However, the approach of indigenous peoples evident in the ecology of their territories has made it possible to understand that their forests are in fact a mosaic of different ecosystems that have been . . . managed for centuries. . . . Indigenous peoples have managed to obtain from the forests . . . all of their needs . . . without destroying these ecosystems. This type of intervention has been called the *anthropogenic forest* or *humanized forest*.

Painfully evident by the late 1990s was the way such an agroecological gradient is being disrupted by the massive deforestation wrought by outsiders. Instead of utilizing an intact forest with small agropastoral openings within it, recent immigrants to the Rio Caquetá—mostly peasants fleeing violence elsewhere in war-torn Colombia who take any wage work they can get—had cut large clearings for pasturing hundreds of cattle; the maize and coca fields that they established were also larger than the typical Ingano fields and orchards. There was also a sharper break between forest and field in the immigrants' colonies than in the traditional harvesting areas shaped by the Ingano. And, of course, the colonists had often encroached on Ingano lands, chainsawing away the forest and forcing some of the Indians to be their sharecroppers for drug production. The colonists might eventually move on, but the damage would already be done. As a result, the Ingano shamans known as *taitas* had to travel deeper and deeper into the forests in order to obtain their traditional medicinal, culinary, and spiritual herbs. The rest of the community often stayed behind, on the edge of a wounded forest, trying to farm where soils had already been depleted.

The Ingano health promoter Libia Diaz has noted that once her people's humanized forests are fragmented, degraded, or eliminated, many of the plants that formerly ensured Ingano health become less accessible. There are four food plants that the Ingano feel are essential to their health, but one in particular, *yoco*, has become rather scarce. Altogether, eighteen declining food species have been targeted by the Ingano for recovery and reincorporation into their contemporary diet on the basis of the following criteria: their traditional (historic) importance; their nutritive value; whether they meet cultural necessities; ability to improve their availability; the enjoyment shared by harvesting, processing, and eating them; and their affordability if purchased from neighbors. A few of the targets are domesticated species, but most are "managed" resources found in the most healthy and diverse of the humanized forests of Colombia's Amazonian headwaters.

Ironically, this region intrigued Vavilov, even though he only reached its edges. When he began his efforts to map the important centers for crop domestication and diversity in 1926, he knew too little of South America to include much detail. But in his revision of the map of "Vavilov centers" in

1928, he roughed out the southern Colombia highlands and piedmont as part of a larger Andean center. As he learned more about South America from Russian colleagues such as Bukasov, he shaped the boundaries of the South American center on the basis of what was known about tuber distributions at the time. Then, in his last version of the Vavilov centers, mapped in 1940, he used multiple criteria—including data from his own travels—and listed Colombia as one of three subcenters in South America.

More recent studies of a larger set of crops have prompted other revisions of Vavilov's map. One of his last foreign students, botanist J. G. Hawkes, maintains that the southern reaches of Colombia, including the Cauca and Caquetá, fall within Vavilov's concept of a center of crop diversity. The World Wildlife Fund has featured the moist tropical forests of the Caquetá in its priorities for critically important terrestrial ecoregions harboring high biodiversity.

Although plant geographers and biogeographers since Vavilov's time have acknowledged the role of indigenous residents such as the Ingano in maintaining and locally enhancing the biodiversity of rain forests, few of the international conservation organizations working in the Amazon actually engage directly with indigenous communities in planning and comanaging the diversity of those habitats. Some, in fact, have attempted to buy land in indigenous homelands to establish protected areas that would exclude agriculture of any kind. That is why the long-term relationships forged between the Ingano and smaller, more dynamic organizations such as the Amazon Conservation Team are so disproportionately important to the future of the conservation of biodiversity. They not only recognize the "humanized forests" in our midst, but also engage the original stewards of those habitats in maintaining their nutritional, medicinal, and ceremonial uses.

Malnutrition certainly exists in indigenous communities in the Colombian Amazon that have been affected by rapid economic and environmental changes, but there has never been the kind of all-out famine in the rain forests that Vavilov witnessed on the collective farms of the semiarid plains and temperate forests of the Soviet Union in 1933.

The 160-some food species found in the homelands of the Ingano will buffer them from starvation just as long as the humanized forests continue to exist. Although they existed in an altogether different landscape and society, the Soviet collectives never fostered such diversity on the ground, even though Vavilov had made it available on the 155 experimental farms found throughout the USSR. There was a tragic disconnect between the world-class food and agricultural research done by Vavilov's empire and its application in solving the problems persisting in Soviet agriculture after the revolution. Although many of the hardy crop varieties that Vavilov discovered did make it out into the fields, gardens, and orchards of the Soviet masses, the collectives were not structured in ways that provided incentives for the farmers to take full advantage of those food resources. Social and emotional issues regarding the organization of farm labor and the confiscation of private property aggravated the famine, regardless of the number of crop varieties the collectives had planted. And there was the inevitable time lag between evaluating which crop resources were best suited to production in particular regions and getting them out to the farmers who needed them most.

When Vavilov returned from the Americas to his institute in Saint Petersburg in the spring of 1933, he realized that he was being criticized at home for not having saved millions of his countrymen from starvation. For the first time, he saw that one of his own employees, Aleksandr Karpovich Kol', had published a scathing critique of his approach in the *Economischeskaia zhihn*, a Communist journal. Serving as the seed curator who coordinated the institute's department of plant introduction, Kol' was ten years older than Vavilov and jealous of the younger man's reputation as a scholar. At the same time, he had been reprimanded by Vavilov so many times for losing or mislabeling valuable seed collections brought back from overseas that Vavilov had finally demoted him. Kol' retaliated by waiting until Vavilov was off to America before he released his critique. He complained that Vavilov was too theoretical in his approach to agricultural problems and too neglectful of the need to rapidly get the best crop plants into the conduit for food production. Kol' even attempted to smear Vavilov as an enemy of the working classes.

Vavilov immediately published a rebuttal to Kol' in the same newspaper, revealing that Kol' had attempted to undermine him ever since being demoted for chronic laziness and scientific sloppiness. For a brief moment, Vavilov's explanation was accepted as the "end of the debate" by most readers. But the criticism soon resurfaced and spread like spilled acid among bureaucrats and his more dissatisfied students. As the famine of 1933 proceeded in killing millions, Soviet politicians as well as scientists seriously began to wonder whether there was a quicker way to bring food security to the world than the grand scheme that Vavilov had envisioned through his world collection of plant genetic resources.

Vavilov was losing political ground at home, but his insights into the rain forests of South America remind us of something too often forgotten: The forest-gardens of the Ingano offer a modicum of food security precisely because they are not so overly managed that the wild foods and medicines essential to community health are kept out of the picture. The collective farms in the Soviet Union were undermanaged for diversity in both a social and an ecological sense, and that led to their lack of resilience. As we shall soon see, Stalin's totalitarian vision for the world squeezed the creativity out of his scientists and squeezed the resilience and wildness out of his people's farmscapes. His vision for optimizing food production was more like the assembly lines of Henry Ford's automobile factories, while Vavilov's vision for a sustainable agriculture was veering toward that of a wild rain forest.

CHAPTER TWELVE

The Last Expedition

We have come to rely upon a comfortable time-lag of fifty years or a century intervening between the perception that something ought to be done and a serious attempt to do it.

—H. G. Wells

There was, indeed, a time lag between when Vavilov first collected a seed and when its characterization, evaluation, and release to farmers made a difference to the food security of his people. By the time Vavilov stopped traveling abroad, in 1933, he and his colleagues had brought home between 148,000 and 175,000 live seed and tuber samples for the world collection. A decade later, in January of 1943, as he lay dying of starvation, it was clear that his vision for "the planned and rational use of the plant resources found all around the terrestrial reaches of the globe" had not yet reached fruition. As the human and economic costs of World War II set Soviet agricultural research grinding to a halt, only 75,000 of Vavilov's seed samples had been characterized and evaluated; many would be lost or damaged over the course of the war. It took many more years to evaluate the surviving samples, and then they trickled out into farmers' fields only slowly.

Be that as it may, Vavilov's promise to the many peoples of the Soviet Republic to feed them with hardy, adapted seed of stocks and tubers gleaned from around the world was not a false promise. Roughly a quarter century after his death, four hundred new crop varieties selected from the seeds he collected were indeed feeding such a large percentage of Soviet citizens that the frequency of famines in the Soviet-controlled republics of Eastern Europe and Central Asia declined precipitously. Early on, some historians hinted that the Soviet populace became more buffered from hunger exactly because of the diversity of food crops that Vavilov and his students had introduced, while others attributed the declines in famine to the government's reorganization of food production units and the modernization of agricultural technologies. It was not until 1979 that Russian food historian and Vavilov biographer G. A. Golubev attempted to assess Vavilov's impact on Soviet food security:

> Four fifths of all the Soviet Union's cultivated areas are sown with varieties of different plants derived from the seeds available in the VIR's unique world collection. The latter has already helped Soviet scientists breed over a thousand valuable varieties known as "Vavilov." [By 1979] these provided the USSR with more than five million tons of additional produce per year, which brought in at least a billion rubles in additional income.

Despite the lack of support for protecting Vavilov's collection in Saint Petersburg during the siege, most of the seeds had been divided into "duplicate" samples and dispersed to various sanctuaries to ensure their survival; the surviving samples became critical to Soviet plant introduction efforts to improve crop yields and food security over the following decades. Of course, Vavilov's contributions to food security should not be assessed merely in terms of gross production. We should consider the enhancement of nutritional quality that occurred as a result of his institute's work, as well. On his last trip to the United States, in 1930, Vavilov became intrigued by the many species of wild sunflowers, which, it seemed, American plant breeders had altogether neglected. On the arid, wind-swept plains of West Texas, Vavilov collected a handful of seeds of *Helianthus lenticularis* to bring back to his staff of oilseed breeders. One of those breeders—V. S. Pustovoit—began a series of handmade crosses between that West Texas sunflower and cultivars of the domesticated sunflower, *Helianthus annuus*, which already produced much of the vegetable oil that Russian Orthodox Christians burned as candles during their Lenten fasts. Even after Vavilov had been dead for decades, Professor Pustovoit persevered in making hybridizations, evaluations, and selections using the wild sunflower that his colleague had brought back.

Thirty years after Vavilov had first given him the sunflower seeds from West Texas, Pustovoit released a stable hybrid sunflower with a level of polyunsaturated oils that dwarfed that of other sunflowers. It became widely grown in the Soviet Union, and in 1972, when farmers from West Texas visited the Krasnodar experiment station on a tour of Soviet agriculture, Pustovoit's daughter offered them seeds of that hybrid. The Texans promised to take it back and cultivate it in the American homeland of its wild ancestors. It was indeed brought into cultivation in West Texas, not far from Vavilov's original collection site. This is but one of many stories that reveals how much the Vavilov legacy led to tangible advances in food production, improved nutrition, and food security, not only in the Soviet republics, but in other countries, as well.

Yet such advances could not be seen—or foreseen—in the aftermath of the 1932–33 drought in the Soviet Union. Suddenly, agricultural researchers were treated as if they had blood on their hands. It was the blood of five million or more who had died on collective *kholkhoz* farms as their grain yields plummeted and their meager harvests were hauled off to the cities, leaving the producers themselves without any food security of their own. The criticism began with vindictive coworkers like Alexsander Kol', who, in 1933, told the OGPU secret police that Vavilov was leading a counterrevolutionary group that operated out of his institute. By that time, the OGPU had already amassed three years of allegations into a file on Vavilov, which as journalist Peter Pringle has amply documented, accused Nikolay "of attempting to sabotage Soviet agriculture." Unlike Lenin, who supported Vavilov's attempts to build a more science-based Soviet food production system, Stalin simply needed a scapegoat for the colossal famine his own policies had brought on. However, Vavilov still seemed too essential to the country's agricultural recovery and too well connected for the OPGU to arrest him at the time. So the question of Vavilov's counterrevolutionary loyalties lay festering like a sore below the skin until 1934, when Trofim Denisovich Lysenko broke it open and turned it into a national—if not international—debate.

Within the decade following 1934, T. D. Lysenko had replaced Vavilov in the role of most powerful scientist in the Soviet republics and had closed down most genetic research and conservation programs in the Union. In essence, by proclaiming that "Marxism is the only true science"—and that other disciplines should be merely subsidiaries in service to it—Lysenko set back Soviet efforts to use science to regain some modicum of food security. Lysenkoism became a "quick fix" ideology that rejected slower, Darwinian plant selection processes as a waste of time, urging that they be replaced by attempts to adapt crops by briefly exposing them to environmental stresses thought to change their genetics. Lysenko and his colleague Michurian built a house of cards on the already discredited theories of Jean-Baptiste Lamarck, who in the eighteenth century expounded on the inheritance of acquired characters without any understanding of genetic processes. Lysenko argued that from William Bateson onward, geneticists had gotten so erudite and elitist in their theories that they no longer worked with plants in the field or directly passed benefits on to "the people." Lysenkoism dragged Soviet biologists and agronomists into the murky backwaters of the life sciences until Lysenko's shoddy experiments and ideological rants were discredited by an overwhelming outcry from other Soviet scientists in 1964.

Lysenko was but one of hundreds of poorly trained but hard-working agronomists and horticulturists involved in crop evaluation and selection in the hinterlands early in his career. However, he had enough practical successes that he was invited to work at the prominent agricultural field station of Odessa, where he met a crafty and scientifically illiterate philosopher named Isai Izrailevich Prezent. As Russian science historian Mark Popovsky darkly admitted,

> The meeting with Prezent changed everything. The cunning . . . philosopher quickly realized the advantage of becoming a mouthpiece of this agronomist who was riding the crest of a wave. . . . Prezent also realized that Lysenko . . . needed something to float on his own program. Prezent set about providing him [with] a philosophical program [based on] Lamarckism. . . . People's Commissariat Yaklovev [had already] demanded that scientists "revolutionize the life of plants ands animals." [Prezent and Lysenko believed that] Lamarck indicated how it could be done.

Lysenko had already established some reputation for novel but rather dubious approaches to selecting plant strains for their adaptations to stress. Bolstered by Prezent's encouragement and full of ideological views that fit comfortably with his superiors' notions of fostering "a people's agriculture," Lysenko became director of the Odessa Institute in 1933, just as the Soviet Union was stunned by new tallies that millions had died from the famine. Lysenko and Prezent promised to make Odessa a stronghold for "barefooted peasant scientists" who would use rustic cabins as their laboratories and common sense infused with Marxist ideals as their science.

With clever phrases offered to him by Prezent, Lysenko gained the support of petty Communist bureaucrats on a "People's Commission," who were increasingly concerned that the educated elite had created a burgeoning research empire that produced few viewable results other than publications. In 1929, even Vavilov tentatively praised as "a major world achievement in plant science" Lysenko's approach to using the masses to "vernalize" or environmentally adapt seed crops. Perhaps he did so only to demonstrate that he could coexist with other approaches and welcome innovative researchers other than his own. However, Vavilov's students cynically joked that Lysenko would claim that his pseudoscience could accomplish any agricultural problem placed before him: "Lysenko is sure that it is possible to produce a camel from a cotton seed and a baobab tree from a hen's egg," they said. While Vavilov certainly recognized that some of Lysenko's claims were based on shoddy science that lacked controls or outside verification of results, he does not seem to have been aware how much Lysenko's style of agricultural research had captured the imaginations of zealous Stalinists.

In May of 1934, Vavilov received a most ominous wake-up call. Stalin had given the Council of the People's Commission the task of determining why the vast agricultural research infrastructure set in place by Lenin and run by Vavilov had not kept millions from starving to death over the previous two years. Why could he not reduce the research and development time needed to release a new high-yielding variety to *two years*, instead of the ten to twelve years Vavilov's team was requiring—and one that most new seed releases still require? Compared to Lysenko's approach, the commissars saw Vavilov's strategies as intrinsically slow and stilted. They separated agricultural research in the academy from the practice of agriculture in the peasants' fields. While Vavilov was in for his equivalent of an annual review, the commissars asked what he would do differently that would avert another such agricultural failure in the years ahead. Couldn't he simply shift his research to document what had happened on the collective farms that had not suffered crop failures during the drought, so as to use them as models for the other collectives?

Vavilov was stunned and left nearly speechless. No one in the room, he believed, really understood science. He had recently been elected by his scientific peers as president of the All-Union Geographical Society, and he had already received nearly every award that scientists around the world could offer to a peer who had advanced their disciplines. About this time, Dr. V. Khmelarzh, one of the most prominent scientists in Eastern Europe, had proclaimed that "the institutes which he [Vavilov] directs are the world's largest and have surpassed American institutes in size and importance." The British geneticist Rowland Biffen—whom Vavilov had idolized as a student—had recently declared that the Soviet Union now occupied first place among all countries in the success of their plant breeding programs. Yet Vavilov was getting little respect at home, especially from the bureaucrats to whom he reported.

When Vavilov finally responded to the commissars, it was not with what they wanted to hear. He intended to stay the course he had been on since the revolution, taking a science-based approach that selected the best breeding stocks for genetic recombination and evaluation under controlled yield trials undertaken in various parts of the country. He would not, as Lysenko had done, rely on untrained farm workers to accomplish the sophisticated agricultural research that would be necessary to stave off future famines.

Within two months, the council released a report on means to avert future famines by restructuring the Union's agricultural research. The council had judged Vavilov's approach to be "utterly insufficient" and detached from what was happening on the best and healthiest of the collective farms. Within the year, Vavilov was demoted, losing power over fifteen of the institutes that had been under his direction, while the budget of his remaining programs was severely cut. By June of 1935, twelve experimental farms were transferred to the State Farm Commissariat, so that Lysenko's model of peasant scientists working in cabin laboratories could be implemented.

In the meantime, Lysenko was receiving direct praise from Stalin. At a 1935 Collective Farmers Conference, Stalin sat in a chair on stage listening to Lysenko vilify the scientific elite, while promoting low-cost quick fixes that he claimed peasant farmers could implement. Stalin rose out of his seat before Lysenko's speech was over, yelling, "Bravo, Comrade Lysenko, bravo!"

Vavilov was also on Stalin's screen, but the subtitle of the image did not include the word "bravo." As late as 1932, Vavilov still had enough prominence and political influence to gain an audience with Stalin to talk him out of keeping two Russian cytologists imprisoned in labor camps. But as the famine dragged on and more people starved, Stalin needed someone to blame. Vavilov knew that he had fallen out of favor with the dictator and began to realize that he was under surveillance. He set up a "code language" in letters to his American friend Harry Harlan, to signal the outside world that he was in danger.

In the mid-1930s, an event occurred by accident that foreshadowed the coming years. Vavilov had the misfortune of nearly running Stalin down as they both turned the corner of a government hallway coming from opposite directions. Although Vavilov offered his apologies, Stalin's response was as cold as ice. Long after Vavilov returned to his office in the institute that day, it was clear to his coworkers that he was suffering from unprecedented stress. The chance encounter with the dictator was all too close and all too untimely to dismiss as trivial. The next time Stalin had an opportunity to encounter Vavilov—at a 1935 conference where both Lysenko and Vavilov were speaking—he first praised Lysenko then left the room just as Vavilov began his address. In a more private meeting, Stalin told Vavilov that scientific expeditions overseas were useless and unsupportable; instead, he said, botanists would do better by spending more time helping farmers secure their harvests: "GO AND LEARN FROM THE SHOCK-WORKERS IN THE FIELDS!" Stalin barked at Vavilov in their last one-on-one conversation.

Vavilov's reaction to Stalin did not take the form of submission. In 1936, when Stalin spearheaded a national campaign to raise the Soviet Union's total grain harvest from 80,000 million to 100,000 million kilograms, Vavilov reminded the Soviet populace that prior to the revolution, farmers had regularly harvested 160,000 to 210,000 million kilos of grain from the same production areas in Russia. This indirect criticism of the dictator won Vavilov an invitation to the Kremlin for an intimate discussion with Stalin's infamous right-hand man, Molotov. Halfway through the "dressing down" by Molotov, Vavilov smelled smoke, then noticed that Stalin had come in through a side door, unannounced, puffing on his pipe. Stalin interrupted the conversation between Molotov and Vavilov to offer three brief sentences before leaving them to work out "the details": "Academician Vavilov, why do you have to have these empty dreams? Just help us to get a dependable harvest of 80,000 million kilos. That's enough for us."

It was clear that Stalin and Molotov did not care about ensuring greater long-term food security for their countrymen; all they wanted was to quell the masses by announcing that they had taken measures that would suddenly increase food production.

That same year, the growing tension between Vavilov and Lysenko came to a head. The scientific adversaries agreed to debate one another before their peers, but more politicians than scientists attended the event. At the onset, Lysenko showed such a lack of knowledge of science that he appeared at times fawning, at times antagonistic, in response to Vavilov's grasp of the underlying scientific issues:

> What I have said in my paper . . . fundamentally contradicts N. I. Vavilov's . . . genetic theory that the corpuscles of "hereditary substance" which remain unchanged over a long chain of generations, create new variation only through recombination. I feel I do not have enough evidence to smash this "law," which I believe contradicts the facts of the evolutionary process of selection. But in the course of my work, I constantly obtain evidence that suggests the unsoundness of this law. . . .

Ironically, evolutionary geneticists since Lysenko's time have indeed found mechanisms for evolution other than sexual recombination of genes that work through gradual processes to the selection of novel forms. That Vavilov understood the gradual processes of evolution as fully as one could in his time but did not immediately discount Michurian's and Lysenko's claims of other possible mechanisms shows him to be a critical but open-minded thinker. We now know there are indeed some mechanisms for more rapid "microevolution," but they are not what Lysenko suggested. The "jumping genes" of Barbara McClintock and the rapid, punctuated evolutionary processes explored by Stephen Jay Gould are but two phenomena that had not yet been recognized at the time of the Lysenko-Vavilov debate. More to the point, Lysenko's contention that he could condition plants and animals so that they rapidly "acquired" new, useful, heritable characteristics through manipulation of their environments is not even in the same league with the work of McClintock and Gould. Lysenko's declaration that Vavilov's methods were "unsound" was based less on scientific evidence than on political ideology. Lysenko realized that he could not defeat Vavilov on scientific grounds alone.

In his reply, Vavilov allowed for the possibility that genetic recombination was not the only possible mechanism that explains the diversity of crops available to humankind:

> Academician T. D. Lysenko is promoting a new hypothesis that the gene is quite variable, and that it can be deliberately changed through the agency of the experimental biologist in pre-determined direction. Unfortunately, as yet there is no experimental evidence that exactly supports this contention. It may well be that in the future, T. D. Lysenko or others may scientifically demonstrate the possibility that such changes occur, and this new science-based knowledge would be welcomed, but as yet, the existence of that process has in no way been proven to us. . . . To disprove the working hypothesis that many geneticists have already developed, we need evidence from precise experiments. These we do not have. . . .

Despite Vavilov's courteous but emphatic dismissal of most of Lysenko's contentions, high-ranking politicians witnessing the debate concluded that "evolution could not have occurred without the inheritance of acquired characteristics." When Vavilov's colleague and Nobel Prize winner Hermann Muller later queried the Communist Party's agricultural secretary about the social implications of that assertion—for it implied that minorities could be considered less genetically advanced than the ruling class—Secretary P. N. Yakolev replied that ethnic minorities were indeed "inferior to us in every respect . . . but after two or three generations of living under the conditions of Socialism, their genes would have so improved [that] we would all be equal."

Lysenko had won the 1936 debate in a political sense, and later that year, it became clear that Vavilov's freedom to move about at will had come to an end. Someone leaked a story to the *New York Times* that Vavilov had been arrested, along with one of his colleagues. In fact, NKVD staffers were still accumulating evidence that they thought might be enough to ensure a death sentence for Vavilov, but they were yet not ready to arrest him. Vavilov had to ask the *Times* to retract the erroneous report and to announce to the world that Soviet science had never been better or more collaborative. Worse yet, he was forced to pose for photos with Lysenko to assure the international community that everything was fine between them.

From 1936 onward, Vavilov worked to set his vision in order, revising his maps of centers of diversity and compiling the geographic data derived from the world collection for a book to be entitled *World Varietal Resources of Grain Crops*. He personally followed some advice that he wrote to a despondent colleague, Konstantin Pangelo: "Work quietly and organize your work [for publication and for posterity] as soon as possible."

Vavilov and Pangelo watched helplessly as more than a dozen and a half prominent geneticists went to the scaffold or disappeared, their crime voicing resistance to Lysenko. Vavilov had long worked to have the Seventh International Congress of Genetics hosted in Moscow in 1937, but Lysenko and Stalin saw to it that the meetings were cancelled. Once it was rescheduled in Edinburgh, Vavilov's international colleagues saw to it that he was elected the honorary president of the Congress; even so, he was not allowed to attend.

Although his days building national institutions and undertaking international expeditions had come to an end, Vavilov retained one role in the scientific community as he had always done: encouraging younger scientists to carry on his work. His enthusiastic mentoring of young men and women did not stop with Soviet citizens such as the Kazakh botanist Aimak Dzangaliev, but extended to people like the American student John Niederhauser, who later won the World Food Award, and the British student J. G. Hawkes, who has brilliantly advanced Vavilov's work on crop geography. Those students—still in their twenties at the time—were invited to dinners at Vavilov's home, taken to the opera with his family, and escorted on botanizing trips into the countryside surrounding Leningrad. Vavilov engaged them in long scientific discussions with his most accomplished understudies, Bukasov, E. E. Leppik, and P. M. Zhukovsky. Tragically, it would soon be realized that it was not in Vavilov's best interests to have personally embraced these British and American citizens, for he would later be accused of spying on behalf of their governments.

By 1939, the writing was on the wall: Genetics would no longer be supported by the Soviet government, since Lysenko had convinced Stalin that he could improve crops by other means. At a March 1939 meeting of the All-Union Institute of Plant Breeding, Vavilov could no longer pretend that he and Lysenko were working toward the same goals. With everyone at the meeting fully cognizant of what he was referring to, Vavilov rose to speak the words that undoubtedly cut short his own life:

> We shall go to the pyre,
> We shall burn,
> But we shall not retreat from our convictions.

From that day on, Vavilov feared for his life nearly every moment, calling his wife both before and after every move he made. Those who visited him in Leningrad saw his crestfallen appearance, as if he had been sick or had gone without sleep for weeks. After more than a year of constant worry for his own safety and for that of his family, he received a call that gave him some hope. His former boss at the People's Commissariat of Agriculture proposed that he make a short expedition to the Carpathian Mountains of the western Ukraine, not far from the border with Poland. Deciding that some fieldwork might help him get his mind off the deplorable state of science in the academy, he accepted the invitation. Russian science journalist Mark Popovsky has described what the chance for one last expedition meant to Vavilov:

> New places, untrodden paths—in his thoughts, he was already breathing the air of the Carpathian mountains and striding through the forests of Bukovina. But this expedition was to be something more than just another excursion to unfamiliar regions. It would serve to release some of the emotional tension that had built up in the institute. [Vavilov] was longing for a change, if only for a time.

On July 23, 1940, Vavilov received a document that authorized him to lead what would be his last expedition. He was fully prepared to make the most of it, having hand-picked the members of the expedition and hand-packed every book and map needed in the field; but then, just before his departure from Moscow, he was called into Lysenko's office. As president of the academy, Lysenko had decided to reject a doctoral thesis on genetics that Vavilov had deemed to be significant work. Their debate over the thesis turned into a larger argument over the fate of genetics and food production in the Soviet Union. The scientific staff working just outside Lysenko's office heard Vavilov fire one last volley at Lysenko just before he slammed the door and rushed away to depart for the Ukraine: "THANKS TO YOU, OUR COUNTRY HAS BEEN OVERTAKEN BY OTHER COUNTRIES."

There was no longer any doubt among the staff present at the academy that day that Vavilov would soon be arrested. Vavilov went by train to Kiev, where he met his staff and members of the Ukrainian Academy of Science on July 26. Within days, his books, plant press, and closest colleagues were all packed in together with Vavilov in a small, black, Soviet-made sedan headed for the mountains. There, in the western Ukraine between Kiev and Lvov, Vavilov had one last chance to participate in the field activities closest to his heart. With his plant-collecting colleagues, he stopped the car every few kilometers to wander through the fields of wheat, barley, oats, and rye, asking the peasant farmers there for permission to take small samples of their seeds to sequester in cloth bags. He wrote up his field notes and the commentaries that those farmers offered him. As the sun went down, he invited the most astute farmers to join him and the other scientists in village cafes and the canteens at the inns where the expedition team would sleep, talking into the wee hours of the night about the problems of agriculture and the promise of certain seeds to solve some of those problems. Vavilov would personally pick up the tab for everyone and go off to his room to sleep, but within a few hours he was back downstairs, still writing up his field notes when the other team members arrived for breakfast at dawn.

On August 6, as Vavilov's team was working in the most remote reaches of the Carpathian Mountains between Chernovitsy and the borders with Poland and Romania, his dust-covered black sedan was intercepted by another of the same model and color. Nikolay himself was not near the roadside, but had begun to hike with a few from his crew up a mountain trail, collecting wild grasses that he thought might be related to spelt or emmer wheats. Oddly, most of Nikolay's colleagues had lingered behind, saying that they wished to enjoy the warm, sunny day and the camaraderie that had developed among them. These coworkers were the ones who heard a question rudely hurled at them from the half-open window of the second black sedan:

"Where is Academician Vavilov? We must find Academician Vavilov!"

His colleagues were told that Vavilov was needed to present important documents on grain exports to the Commissariat of Agriculture. This puzzled Vavilov's younger coworkers, like botanist Vadim Lekhnovich, who could not imagine why a bureaucrat was shouting for his boss to present papers in the midst of a remote field site. When Vavilov returned from his hike, he was suddenly taken away by the four men in the second black sedan. His crew was told that Vavilov had been called back to Moscow on business, but that they should continue with their plant explorations. Young Lekhnovich was given a note in Vavilov's hand that simply said, "In view of my sudden recall to Moscow, hand over all my things to the bearer of this note. N. Vavilov, August 6, 1940, 2315 hours."

Lekhnovich and the others did what they were told, taking every scrap of paper, cloth, and instrument belonging to Vavilov out to the second black sedan, which had returned to the hotel in Chervotsky. When the car disappeared, that was the last any of the scientists would see of Vavilov or his personal possessions.

Except Vavilov's rucksack, which mysteriously reappeared at VIR in Saint Petersburg some time later. In it was an ancient, perhaps wild form of wheat, with primitive traits found in both the spelt and emmer subspecies.

"What a find!" his Russian colleagues exclaimed, because just that week Vavilov had been challenging all the Ukrainian scientists he had met to be on the lookout for what might remain of the original forms of wheat brought in through Balkan countries thousands of years earlier. Vavilov had apparently beat them to the punch, discovering a most primitive form of wheat on his own. It was his last collection, and years later it was determined to be a new species of wheat in the genus *Triticum*.

In the two years that followed, Nikolay Vavilov suffered through thousands of hours of interrogation, sometimes for ten to thirteen hours a day while standing up. He was accused of making plans to escape from the Carpathian Mountains into Poland and of spying for Great Britain and for anti-Soviet organizations. But the most horrific accusation was that he had intentionally "wrecked" Soviet agriculture in the 1930s by mandating that so much new acreage be cultivated in grains that there was not enough seed for adequate planting. As a result, the court concluded, the country's fields in 1931 and 1932 were so short of crop seedlings and so full of weeds that a massive famine was precipitated. Stalin had at last secured his scapegoat for the famine that had killed millions.

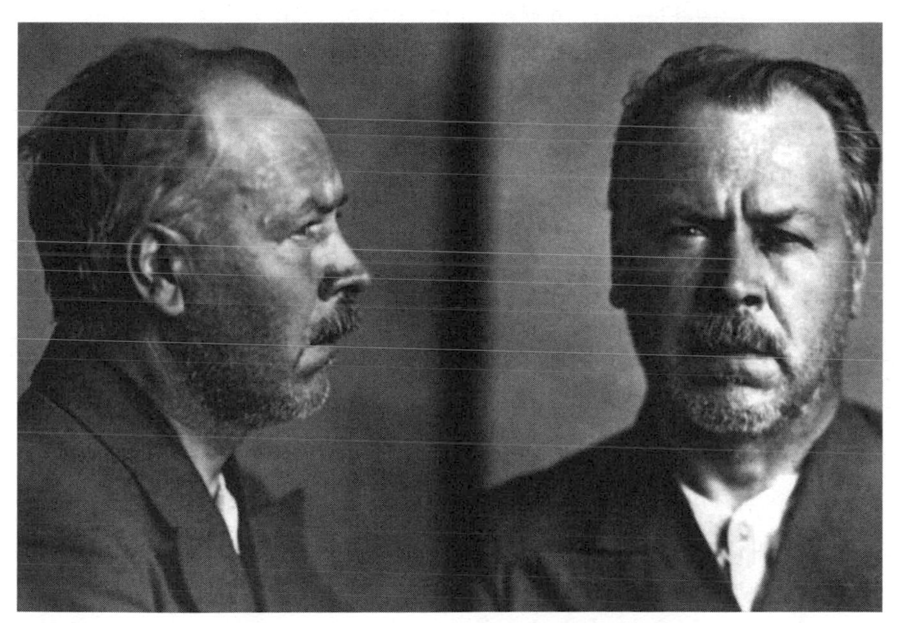

Nikolay Vavilov, in prison, probably taken at Saratov in 1941.

On July 9, 1941, a closed session of the Military Collegium of the Supreme Court of the USSR brought Vavilov out of his cell in a Moscow prison for a few minutes, to find him guilty of all offenses. The court declared that "Nikolay Ivanovich should suffer the supreme penalty—to be shot and to have all his property confiscated."

But Vavilov was not shot; others had intervened to commute his death sentence and to petition for his release. Most of his former colleagues, however, had no idea what had happened to him. In the news section of the scientific journal *Chronica Botanica*, it was first reported in 1941 that Vavilov was arrested and was being held in custody; a year later, it was reported that he had died in Siberia. Yet the organization Russian War Relief Incorporated dispatched a press release in 1942 that claimed, "Trofim Lysenko and Nikolay Vavilov, two leading Russian agricultural scientists—known as Russia's present day Burbanks—are both devoting their efforts in war-time to increasing the Russian food supply . . . through the spread of scientific farming . . . [putting aside] their sharp disagreements on problems of theory and practice . . . to instruct farmers in ways of increasing their yields."

Ironically, late in 1941, Vavilov was transferred to a prison in Saratov, the very town where he had begun his distinguished agricultural research career. There, finding many scientific colleagues who had also been purged by Stalin, he offered hundreds of hours of scientific lectures to his fellow inmates. He thought that he would soon be allowed to practice his science once more, pledging to his interrogators that his only wish was to use his talents to improve the food production and nutritional welfare of his countrymen. But by the spring of 1942, it became clear that he and other inmates were slowly being starved to death. Most of them suffered from acute dysentery, and a madman was placed in Vavilov's cell who regularly stole his ration of bread. His health and his hope gradually deteriorated.

On January 26, 1943, Nikolay Ivanovich Vavilov—the man who more than anyone else in history helped humankind appreciate where our food has come from—died from the side-effects of slow starvation.

Epilogue

"The hope—from Vavilov's time to the present—has been that by saving seed diversity in collections now known as gene banks, humankind will have a buffer against famine caused by plagues, pestilence, floods, and other catastrophes, in a manner that staves off hunger. But by the year 2000, between 786 and 797 million people in the world were considered to be chronically hungry, depending on when exactly the counting was done. Another 2 billion people were said to suffer from "hidden hunger," that is, a chronic deficiency in iron, vitamin A, or other essential nutrients.

It is a sad irony that half of all individuals labeled as "the hungry" are farmers, small shareholders of lands that are likely to be located in those centers of food diversity that Vavilov first described and places like them. Another 8 percent of the hungry work in food production as herders, fishers, foragers, or hunters, but do not gain enough income or edible biomass from those activities to stave off starvation. Some 22 percent of the hungry are landless in rural areas, whereas 20 percent lack access to food-producing lands inside the boundaries of cities.

In a world where there are roughly as many overnourished people as there are undernourished people at risk of starvation or disease, the causes of hunger and food security problems can no longer be seen simply as scarcity of adapted seed varieties or the underproduction of nutritious foods. Unequal access is the underlying dilemma.

"The root cause of hunger isn't the scarcity of food or land; it's the scarcity of democracy," said Frances Moore Lappé. As she has so many times since writing *Diet for a Small Planet* some four decades ago, Lappé cuts to the chase in *World Hunger: Twelve Myths,* which she coauthored with Joseph Collins and Peter Rosset.

Stalin's henchmen blamed the 1931–32 famine on seed scarcity, which they claimed had been precipitated by Vavilov's long-term strategies and science policies. But the scarcity of seeds—or, better, the depletion of seed diversity—is seldom even the proximate cause of famine, and when it is a factor, the ultimate cause of hunger is largely political. More to the point, it is the *social, economic, and political access to seed diversity* at critical moments that can make or break a community's means of achieving food security. We do not know whether Nikolay Vavilov acknowledged this himself until he was close to his death bed, when he fully realized how much his attempts to deal directly with food security issues had been sabotaged by scientific adversaries with personal motives.

Of course, there is another factor to consider: oppressive governments. Is it any surprise that the incredible seed diversity championed, conserved, and studied by Vavilov did not bear as much fruit as it otherwise could have, given that he ended his life and his career in the totalitarian state that Stalin so insidiously crafted? How can diversity flourish under any form of totalitarianism? Whether that totalitarianism is one that is fostered by a blind faith in communism, fascism, religious fundamentalism, or capitalism (and its corollary, the primacy of private property rights), it leaves little room for the proliferation of a *food democracy*. In such a democracy, all citizens can choose how to practice their right to feed themselves and their families an adequate supply of healthful, nutrient-rich, toxin-free, culturally appropriate foods.

Despite the remarkable efforts made by Vavilov, Stalin's political regime and hierarchical structure ultimately undermined many of his contributions. Quite simply, Josef Stalin and his Secretary of the Academy of Sciences, Nikolay Gorbunev, scapegoated Vavilov when their own collective farm program fell into ruins. The peasant farmers forced onto local and state collective farms were forced to grow particular foods, such as approved bread wheat varieties for the state, and most of their harvest went to the cities. At the same time, many rural families were discouraged or prohibited from sowing a diversity of crops on their former landholdings as a means to feed their families. The concept of food democracy as we articulate it today would have seemed painfully remote to them.

Agricultural law scholar and activist Neil Hamilton uses the term *food democracy* in a rather precise manner:

> The word "democracy" comes from Greek words meaning "people" and "rule." How, then, do we make the people rule our food system? There are four essential pieces in the creation of a food democracy. The first is citizen participation; all actors in the food system must have a voice, and the contributions and concerns of each group must be considered. Second, informed choices are necessary. Questions, information, and knowledge about how food is produced are key. Third, a number of choices must be available to citizens. Although there are currently many types of food to choose from, most of the food is produced in the same faceless, industrial manner. Fourth, participation in food democracy must happen at the local as well as the national levels. One's food choices should be geared toward protection and development of the community, whether this means buying from farmer's markets or eating at locally owned restaurants.

By Hamilton's definition, Vavilov helped people in the Soviet Union meet two of the conditions for food democracy—more diverse choices of what to grow, and better scientific information on which to base their economic and agronomic decisions—but he had little or no effect on their capacity to participate in decision making at either the local or the national level. If anything, his reliance on a centralized, rather hierarchical structure for collecting, storing, evaluating, and disseminating seeds made active participation of the populace difficult.

That may be the only legitimate reason that Stalin could have offered for his refusal to give the institute's seed bank the protection and support it deserved, even as he offered such support to the art collections in the Hermitage. The Russian art curators could claim that they were saving the art "for the future enjoyment of all people," that they had taken those treasures of our common heritage out of the hands of the czarist elite and were keeping them from falling into the hands of the German elite. Stalin was not perceptive enough to recognize that the seeds could also be considered resources of our common heritage or that seed banks were merely meant to be a backup in case peasant farmers lost their own seeds to weather, pestilence and plague, or war.

Instead, it was the tenacity of Vavilov's loyal staff at the All-Union Institute of Applied Botany and New Crops that kept the seeds alive, even though the vaults and field plots that harbored the World Collection of Cultivated Plants suffered through two decades more of Lysenko's despotic control after Vavilov's death. As Nikolay Vavilov's friend P. M. Zhukovsky reported in 1968, many accessions of the seeds "being long subjected to cross-pollination, out-breeding and introgression . . . have lost their authenticity. . . . During the World War, civil wars and [various military] occupations, many samples have lost their germinability."

Zhukovsky's comments reveal the vulnerability that many national seed banks have historically faced during civil strife, but let us make no mistake in thinking that the era of destroying seed banks—by warfare or by plain neglect—is a thing of the past. In 1992, the mujahideen fighters completely destroyed the national seed collection of Afghanistan. Concerned Afghani scientists re-collected some of the seeds and sequestered them within private homes in two cities, but those, too, were destroyed when further warring between the Taliban and their adversaries destroyed the neighborhoods where the seeds were secretly harbored.

And in 2003, while the world media agonized over the looting of the Iraqi National Museum of some fifteen thousand antiquities during the chaos generated by the American invasion of Baghdad, few journalists gave any attention to the destruction of Iraq's main seed bank and plant breeding program a few kilometers away at Abu Ghraib. Yes, at the same Abu Ghraib where torture of Iraqi prisoners later occurred, Iraq had harbored a gene bank with some of the finest collections of ancient Mesopotamian seeds that had survived through the twentieth century. In a situation eerily parallel to the Blokada that had occurred in Leningrad during World War II, while all were focused on the looting of art and antiquities collections of the country, the gene bank was intentionally dismantled and left in ruins.

Fortunately, a few Iraqi scientists who worked at the seed bank had already boxed up duplicate copies of the unique and most valuable seeds from the bank. They had sent those priceless seeds off to colleagues in Syria well before the American and British invasion of their country began. A few months after the invasion occurred, science journalist Fred Pearce was allowed a glimpse of that treasure:

> It was just a battered old brown cardboard box sealed with tape, pulled from a high shelf in a refrigerated seed bank in Syria. But this, I was assured, was the famed "black box"—the genetic holy grail, the ark of the lost seeds, the future agricultural prosperity of Iraq.
>
> The box was put together in 1996 in the Baghdad suburb of Abu Ghraib. Known mainly for its notorious prison, Abu Ghraib was once the home of Iraq's main seed bank and plant breeding program. It was here that plant scientists, fearing for the future of their collection, packed up more than 1000 vital seed varieties—everything from ancient wheats to chickpeas, lentils and fruits—and shipped them off to Aleppo for safe-keeping.

To buffer the great seed banks of the world from potential calamities of this sort—caused by political strife or natural disaster—the Global Crop

Diversity Trust was established as a public institution in October of 2004. Its specific task is to help safeguard the collections of some 1,500 national and international seed banks around the world. Many of those seed banks have had no secure long-term funding and have not made duplicate collections for safeguarding in other like-minded institutions, should disasters happen on their own turf. Cary Fowler, director, sees the Global Crop Diversity Trust as an effort to reconstitute and "complete" the world collection that Vavilov once envisioned. Fowler and his colleagues are raising an endowment of $260 million to guarantee the *ex situ* (gene bank) conservation of the seed and tuber resources of some thirty-six food crops essential to global food security, as well as some eighty species of forage crops. Such an endowment could bring management security to hundreds of thousands of plant varieties already in collections that are not currently given all the care they deserve. The trust is linking efforts to strengthen management in historically important seed banks with the already famous initiative to place seeds in backup storage in a frozen vault within the Arctic Circle, the Svalbard Global Seed Vault.

While those efforts are laudable in providing an international safety net for some of the world's most widely distributed food crops, there is another, more difficult and culturally complex, task that needs to be completed if more than just plant breeders are to participate in a food democracy of global proportions. The gene banks backstopped by the Global Crop Diversity Trust have pledged to democratize access to seeds to be used for crop improvement and seed bank repatriation purposes, rather than excluding some political adversaries from access or allowing patenting of those materials of fundamental importance to humankind. Nevertheless, seed banks are not designed to maintain the ongoing processes of farmer-based (in situ) seed conservation and exchange. Their seeds are "frozen in time"—that is, they are more or less adapted to conditions in the year of their last field grow-out—and are not as dynamically changing in response to global climate change, disease, and pest introductions as are seed crops annually grown out in field settings. Moreover, the corporate and academic plant breeders who are the most common recipients of seeds from those repositories typically do work that is a poor substitute for that done on-farm by "vernacular plant breeders"—traditional farmers.

The time-tried strategies of crop selection and diversification that traditional farmers have used in Vavilov's centers of food biodiversity provide us with most of the food we eat. The very processes of farmer-based selection and adaptation that have propelled crop evolution continue in farmers' fields today, and at least in some places, are even keeping pace with climate change, pests, and diseases. We must honor and support those farmers just as much as we support seed banks, for they do essential frontline work, just as the seed banks serve as essential backup. For that reason, this book has focused on Vavilov's attention to how farmers keep and exchange seed, fruit, and tuber varieties in the so-called centers of diversity, and not merely on how he assembled the world's most comprehensive gene bank.

As agricultural policy experts Andrew Mushita and Carol Thompson have recently affirmed, "The future of the planet depends not so much on military power nor on capital speculation but on each one of us making daily food choices that affect global exchange or private enclosure of biodiversity—our collective nourishment, our wealth."

An investment commensurate to that being made for the Global Crop Diversity Trust should also be made to support indigenous and other rural cultures in each of the so-called Vavilov centers, so that on-farm diversity can be maintained for the purposes of local and regional food security. Such on-farm conservation like that being done by Quechuan farmers in the Parque de la Papa in Peru is one model for sustaining food diversity. Effectively generating and dynamically adapting on-farm diversity to changing conditions are among all farmers' fundamental strategies for responding to the many climatic, ecological, biotic, economic, and political shifts occurring in their midst.

As we have seen in the previous stops along Vavilov's journey, farmers everywhere are faced with global warming; international market pressures; introduced pests, weeds, and diseases; and depleted soils and aquifers; as well as increasing frequencies of earthquakes, hurricanes, floods, and other natural disasters. Rather than being passive victims to such changes, many farmers use their seed diversity, their traditional ecological knowledge, and a variety of technologies to adapt with resilience to shifting conditions. But environmental and economic conditions are now shifting faster than ever before, such that farmers may need more outside support and technical consultation than their forefathers required.

Because seed diversity is one of the most effective means of buffering themselves against detrimental change, farmers, gardeners, and orchard-keepers around the world are now affirming their "farmers rights" as one means to remain engaged with the many food resources developed by their ancestors, their contemporaries, and themselves. They do not necessarily want to turn over their seed resources to a multinational corporation, a government bureaucracy, or even a network as benign as the Global Crop Diversity Trust. Rather, they wish to be in active partnership with others who are sincerely concerned about the fate of seeds and their role in contributing to food security. It is heartening to see a growing number of grassroots and international organizations taking up the challenge of redressing the balance of on-farm conservation and use with backup (long-term) seed conservation, as farmers increasingly reengage in the former.

It will take striking a balance between those two ideals—food democracy and farmers rights—to enable the diverse peoples of the world to fully develop food security. Nikolay Ivanovich Vavilov's pioneering efforts to help us appreciate where our food truly comes from set us on the quest to achieve that balance. It is the work that he left unfinished.

Bibliography

Chapter One: The Art Museum and the Seed Bank

Alexanian, Sergey, and V. I. Krivchenko. 1991. "Vavilov Institute scientists heroically preserve world plant genetic resources collections during World War II Siege of Leningrad." *Diversity* 7(4): 45–50.

Chapin, Mac. 2004. "A challenge to conservationists: Can we protect natural habitats without abusing the people who live in them?" *World Watch* 17(6): 17–30.

Cohen, Barry Mendel. 1980. "Nicolai Ivanovich Vavilov: His life and work." PhD dissertation, University of Texas, Austin.

Dorofeev, V. F., and A. A. Filatenko. 1987. "Establishment and development of N. I. Vavilov's theory about the centers of origin of cultivated plants." *Soviet Genetics* 23(11): 1340–53.

Ellis, Neena. 1995. "Vavilov's ghost." Laurel, MD: Soundprint Media Center for National Public Radio.

Golubev, G. 1987. *Nikolai Vavilov: The Great Sower—Pages from the Life of the Scientist.* Moscow: Mir Publishers. Originally published in Russian in 1979, translated into English by Vadim Sternik.

Harris, David R. 1990. "Vavilov's concept of centers of origin of cultivated plants: Its genesis and its influence on the study of agricultural origins." *Biological Journal of the Linnaean Society* 30: 7–16.

Leppik, E. E. 1969. "Life and work of N. I. Vavilov." *Economic Botany* 23(2): 128.

Nabhan, Gary Paul, and Carol Thompson. 2006. "Status and trends in agricultural biodiversity." Draft (unpublished) report commissioned for the UN Food and Agriculture Organization, Rome, Italy.

Popovsky, Mark. 1984. *The Vavilov Affair*. North Haven, CT: Archon Books.

Pringle, Peter. 2003. *Food, Inc.: Mendel to Monsanto—The Promises and Perils of the Biotech Harvest*. New York: Simon and Schuster.

Pringle, Peter. 2008. *The Murder of Nikolai Vavilov*. New York: Simon & Schuster.

Reznik, Semyon. 1968. *Nikolai Vavilov*. Moscow: Molodaia Gvardia.

Reznik, Semyon, and Yuri Vavilov. 1997. "The Russian scientist Nikolay Vavilov." In *Nikolay Ivanovich Vavilov: Five Continents*, edited by L. E. Rodin, translated from the Russian by Doris Love, xviii–xxviii. Rome: IPGRI (Bioversity International); St. Petersburg: VIR.

Rodriguez, J. P., et al. 2007. "Globalization of conservation: A view from the South." *Science* 317(5839): 755–56.

Rodriguez-Armesto, Felipe. 2003. *Near a Thousand Tables: A History of Food*. Madrid: Pan Books.

State Hermitage Museum. 2006. "Hermitage History." http://monarch.hermitage.ru/html.

St. Petersburg Press. 2005. "Art Heritage Saved for Humanity. www.museum-security.org/petersburg2.html.

Stroebel, Gabrielle. 1993. "Seeds in need: The Vavilov Institute." *Science News*, December 18–23, pp. 2–6.

Varshavsky, Sergei P., and Boris Rest. 1985. *Saved for Humanity: The Hermitage during the Siege of Leningrad, 1941–1944*. Leningrad: Aurora Art Publishers.

Vavilov, Nikolay Ivanovich. 1935. "Phytogeographic basis of plant breeding" and "Theoretical basis of plant breeding." Translated and reprinted as pp. 13–54 in K. Starr Chester, ed., 1949–50, "The origin, variation, immunity and breeding of cultivated pants: Selected writings of N. I. Vavilov." *Chronica Botanica* 13 (1–6).

———. 1997. *Five Continents*. Edited by L. E. Rodin, translated from the Russian by Doris Love. Rome: IPGRI (Bioversity International); St. Petersburg: VIR.

Chapter Two: The Hunger Artist and the Horn of Plenty

Aronson, T. 1973. *Grandmama of Europe: The Crowned Descendants of Queen Victoria*. London: Cassell.

"Biodiversity for food security." 2004. Statement by FAO and CTA, World Food Day, October 16, 2004. www.cta.int/about/biodiv_statement.htm.

Brooks, F. T. 1927. "Disease resistance in plants." *Phi* 27(2): 85–87.

Cohen, Barry Mendel. 1980. "Nicolai Ivanovich Vavilov: His life and work." PhD dissertation, University of Texas, Austin.

Cutler, P. 1984. "Famine in Ethiopia." *Disasters* 16.

Dando, W. A. 1980. *The Geography of Famine*. New York: Halsted Press.

Edgar, W. 1892. "Russia's conflict with hunger." *American Review of Reviews*.

Golubev, G. 1987. *Nikolai Vavilov: The Great Sower—Pages from the Life of the Scientist*. Moscow: Mir Publishers. Originally published in Russian in 1979, translated into English by Vadim Sternik.

Golubev, Genady, and Nikolai Dronin. 2004. "Geography of droughts and food problems in Russia (1900–2000)." Report of the International Project on Global Environmental Change and Its Threat to Food and Water Security in Russia. Moscow.

Kendrick, John M. L. 1998. "The Spala Crisis of 1912: Rasputin explained and Alexei found." www.npsnet.com/alexei_found/.

Lanin, E. 1891. "Famine in Russia." *Fortnightly Review* 56: 640.

Lilly, David P. 1995. "The Russian famine of 1891–92." Department of History Outstanding Paper, Loyola University, Chicago, Illinois. www.artukraine.com/famineart/russfam.htm.

Murton, Brian. 2000. "Famine." In *The Cambridge World History of Food*, edited by Kenneth F. Kiple and Kriemhold Conee Ornelas,1411–27. Cambridge: Cambridge University Press.

Mushita, Andrew, and Carol B. Thompson. 2007. *Biopiracy of Biodiversity: Global Exchange as Enclosure*. Trenton, NJ: Africa World Press.

Popovsky, Mark. 1984. *The Vavilov Affair*. North Haven, CT: Archon Books.

Radziwill, Catherine. 1914. *Behind the Veil of the Russian Court*. New York: Dial Press.

Robins, Robert J. 2002. "Genetics and history: How a single gene mutation affected the entire world." www.esp.org/misc/vignettes/alexis.html.

Rosengrant, Mark W., and Sarah A. Cline. 2003. "Global food security: Challenges and policies." *Science* 302(5652): 1917–19.

Shipton, Parker. 1990. "African famines and food security: Anthropolical perspectives." *Annual Review of Anthropology* 19: 353–94.

Thoreau, Henry David. 1993. *Faith in a Seed*. Washington, DC: Island Press.

Timoshenko, Vladimir P. 1932. *Agricultural Russia and the Wheat Problem*. Food Research Institute Grain Economics Series. Menlo Park, CA: Stanford University.

Twain, Mark. 1869. *The Innocents Abroad, or the New Pilgrim's Progress*. New York: Library of America.

Chapter Three: Melting Glaciers and Waves of Grain: The Pamirs

Conaghan, Eammon. 2004. "Scouting for crop ancestors in volatile outposts." *Ground Cover* 48 (February). www.grdc.com.

Coward, E. Walter. 2006. "Khuf Valley journal." Unpublished manuscript in the archives of the Christensen Fund, Palo Alto, California.

Korzinsky 1893. As quoted in *Nikolay Ivanovich Vavilov: Five Continents*, edited by L. E. Rodin, translated from the Russian by Doris Love, xviii–xxviii. Rome: IPGRI (Bioversity International); St. Petersburg: VIR.

Kurlovich, B. S., S. I. Rep'ev, M. V. Petrova, T. V. Boratseva, L. T. Kartuzova, and T. A. Voluznova. 2000. "The significance of Vavilov's scientific expeditions and ideas for development and use of legume genetic resources." *Plant Genetic Resources Newsletter* 124: 23–32.

Olufsen, O. 1904. *Through the Unknown Pamirs: The Second Danish Pamir Expedition, 1898–1899.* London: William Heinemann.

Pistorius, R. 1977. *Scientists, Plants and Politics: A History of the Plant Genetics Resources Movement.* Rome: IPGRI.

Pringle, Peter. 2008. *The Murder of Nikolai Vavilov.* New York: Simon & Schuster.

Robles Gil, Patricio, ed. 2004. *Hotspots Revisited.* Mexico City: CEMEX.

Street, Ken. 2003. "Following in Vavilov's footsteps." www.new-agri.co.uk.

"Tajikistan 2002: State of the Environment Report." http://enrin.grida.no/htmls/tadjik/soe2001/eng/htmls/climate/state.htm.

Vavilov, Nikolay Ivanovich. 1991. "Near the Pamirs: Agricultural essay." *Bulletin of Applied Botany, Genetics and Plant Breeding* 140: 1–12.

———. 1997. *Five Continents.* IPGRI, Rome, Italy and VIR, St. Petersburg, Russia.

Chapter Four: Drought and the Decline of Variety: The Po Valley

Bazzani, G. M., V. Gallerani, D. Viaggi, M. Raggi, and F. Bartolini. 2004. "The sustainability of irrigated agriculture in Italy under water and agricultural policy scenarios." 9th Joint Conference on Food, Agriculture and the Environment, August 28–September 1, University of Bologna, Department of Agricultural Economics and Engineering. www.tesaf.unipd.it/Minnesota/it/bazzani-et-al.pdf.

"Discovering the Great River: Establishing a new academic model for research into the memory and identity of a region." May 7, 2007. Press release. Pollenzo, Italy: Universitá degli Studi di Scienze Gastronomiche.

Hammer, Karl, Helmut Knüpffer, Pietro Perrino, and Gaetano Laghetti. 1999. *Seeds from the Past: A Catalogue of Crop Germplasm in Central and North Italy*. Bari, Italy: Consiglio Nazionale delle Ricerche, Instituto del Germoplasma.

Hammer, Karl, and Gaetano Laghetti. 2005. "Genetic erosion: Examples from Italy." *Genetic Resources and Crop Evolution* 52(5): 629–34.

Nabhan, Gary Paul. 1992. *Songbirds, Truffles and Wolves: An America Naturalist in Italy*. New York: Pantheon/Penguin Books.

Negri, Valeria. 2003. "Landraces in central Italy: Where and why they are conserved and perspectives on their on-farm conservation." *Genetic Resources and Crop Evolution* 50: 871–85.

Negri, Valeria, and Nicola Tosti. 2002. "*Phaseolus* genetic diversity maintained on-farm in central Italy." *Genetic Resources and Crop Evolution* 49: 511–20.

Pringle, Peter. 2008. *The Murder of Nikolai Vavilov.* New York: Simon & Schuster.

Robles Gil, Patricio, ed. 2004. *Hotspots Revisited*. Mexico City: CEMEX.
Soldati, Mario. 1957. *Travel in the Po Valley. Su Maesta il Po*. Verona, Italy: A. Mondadori. New edition, 1984.
Vavilov, Nikolay Ivanovich. 1997. *Five Continents*. Edited by L. E. Rodin, translated by Doris Love. Rome: IPGRI (Bioversity International); St. Petersburg: VIR.
Zwingle, Erla. 2002. "Po: River of pain and plenty." *National Geographic* (May).

Chapter Five: From Breadbasket to Basket Case: The Levant

Al-Rihani, Amin. 1965. *Qalb Lubnan*. Beirut: Dar al-Rihani.
Kilani, Hala. 2007. "*Al Hima* revives traditional methods of conservation and poverty reduction." World Conservation Union. www.iucn.org/places/wescana/news/hima_workshop.html.
Massad, Barbara Abdeni. 2005. *Man'oushé: Inside the Street Corner Lebanese Bakery*. London: Alarm Books.
Orfalea, Gregory. 2006. *The Arab Americans: A History*. Northhampton, MA: Olive Branch Press/Interlink Publishing.
Toufeili, I., A. Olabi, S. Shadarevian, M. Abi-Antoun, R. Zuryak, and R. Baalbaki. 1997. "Relationships of selected wheat parameters to burghul making quality." *Journal of Food Quality* 20(3): 211–24.
Traboulsi, Fawzaz. 2007. *A History of Modern Lebanon*. London: Pluto Press.
Vavilov, Nikolay Ivanovich. 1997. *Five Continents*. Edited by L. E. Rodin, translated from the Russian by Doris Love. Rome: IPGRI (Bioversity International); St. Petersburg: VIR.
Yammin, al-Khuri Antun. 1919. *Lubnan Ba'd al-Harb, 1914–1919*. Beirut: Al-Matba'ah al-Adabiya.
Zuryak, Rami. 2007. "The roots of inequality." *El-Tayeb Quarterly Newsletter* (April), p. 6.

Chapter Six: Date Palm Oases and Desert Crops: The Maghreb

Belgrave, C. Dalrymple. 1924. *Siwa: The Oasis of Jupiter Ammon*. London: John Lane.
Fakhry, Ahmed. 1974. *The Oases of Egypt: Siwa Oasis*. Cairo: American University of Cairo Press.
Forbes, Robert Humphrey. 1921. "Siwa oasis." Crop Science Society oral presentation notes. Manuscript on file in the Forbes archival collection, Arizona Historical Society, Tucson.
Nabhan, Gary Paul. 2007. "Agrobiodiversity change in a Saharan Desert oasis, 1919–2006: Historic shifts in Tasiwit (Berber) and Bedouin crop inventories of Siwa, Egypt." *Economic Botany* 61(1): 31–43.
Vavilov, Nikolay Ivanovich. 1997. *Five Continents*. Rome: IPGRI (Bioversity International); St. Petersburg: VIR.

Walker, Brian, and David Salt. 2006. *Resilience Thinking: Sustaining Ecosystems and People in a Changing World*. Washington, DC: Island Press.

Chapter Seven: Finding Food in Famine's Wake: Ethiopia

Andenow, Y., M. Hullakal, G. Belay, and T. Tesemma. 1997. "Resistance and tolerance to leaf rust in Ethiopian tetraploid wheat races." *Plant Breeding* 116(6): 533–36.

Assefa, Kebebew, Hailu Tefera, Arnulf Merker, Tiruneh Kefyalew, and Fufa Hundera. 2001. "Quantitative trait diversity in tef [*Eragrostis tef* (Zucc.) Trotter] germplasm from Central and Northern Ethiopia." *Genetic Resources and Crop Evolution* 48: 53–61.

Chossudovsky, Michel. 2000. "Sowing the seeds of famine in Ethiopia." *The Ecologist* (October 1). Reprinted in *Global Research*, September 10.

Clark, J. D., and M. A. J. Williams. 1978. "Recent archaeological research in southeastern Ethiopia." *Annales d'Ethiopie* 11: 19–44.

Demissie, Abebe. 1991. "Potentially valuable crop plants in a Vavilovian centre of diversity: Ethiopia." In *Crop Genetic Resources of Africa*, edited by F. Attere, H. Zedan, N. Q. Ng, and P. Perrino, vol. 1, pp. 89–98. Rome: IPGRI.

Edwards, S. B. 1991. "Crops with wild relatives found in Ethiopia." In *Plant Genetic Resources of Ethiopia*, edited by Jan N. N. Engels, J. G. Hawkes, and Melaku Worede, pp. 42–74. New York: Cambridge University Press.

Engels, Jan N. N., J. G. Hawkes, and Melaku Worede. 1991. *Plant Genetic Resources of Ethiopia*. New York: Cambridge University Press.

Expert Panel on the Stem Rust Outbreak in Eastern Africa. 2005. "Sounding the Alarm on Global Stem Rust." Mexico: CIMMYT, Mexico City. www.cimmyt.org/English/wps/news/2005/aug/pdf/Expert-Panel-Report.pdf.

Kebebew, Fassil, ed. 2001. *An Action Plan for Biodiversity Conservation in Ethiopia*. Addis Ababa: Ethiopian Institute for Biodiversity Conservation.

Kebebew, Fassil, Yemane Tsehaye, and Tom McNeilly. 2001. "Morphological and farmers cognitive diversity of barley (*Hordeum vulgare* L.) [Poaceae] at Bale and North Shewa of Ethiopia." *Genetic Resources and Crop Evolution* 48: 467–81.

MacKenzie, Debora. 2007. "Billions at risk from wheat super-blight." *New Scientist* 2598 (April 3): 6–7.

Raloff, Janet. 2005. "Wheat warning: New rust could spread like wildfire." *Science News* 168(13).

Pringle, Peter. 2008. *The Murder of Nikolai Vavilov*. New York: Simon & Schuster.

Shames, Seth. 2007. *A Financial Analysis of Conservation Tillage as a Solution to Land Degradation on Small Farms in Ethiopia*. Washington, DC: Ecoagriculture Partners.

Stokstad, Eric. 2007. "Deadly wheat rust threatens world's breadbaskets." *Science* 315(5820): 1786–87.

Van Leur, J. A. G., and Hailu Gebre. 2003. "Diversity between some Ethiopian farmers' varieties of barley and within these varieties among seed sources." *Genetic Resources and Crop Evolution* 50: 351–57.

Worede, Melaku. 1991. "An Ethiopian perspective on conservation and utilization of plant genetic resources." In *Plant Genetic Resources of Ethiopia*, edited by Jan N. N. Engels, J. G. Hawkes, and Melaku Worede, pp. 3–21. New York: Cambridge University Press.

Chapter Eight: Apples and Boomtown Growth: Kazakhstan

Browning, Frank. 1998. *Apples: The Story of the Fruit of Temptation*. New York: North Point Press.

Cohen, Barry Mendel. 1980. "Nicolai Ivanovich Vavilov: His life and work." PhD dissertation, University of Texas, Austin.

Dzangaliev, A. D. 1976. 'Fruit forests of Kazakhstan: Their value and usage." Proceedings of the First All-Union Symposium on Man and the Biosphere. UNESCO Man and the Biosphere Program in the USSR, Moscow.

———. 2003. "The wild apple tree of Kazakhstan." *Horticultural Reviews* 29: 63–303.

Dzangaliev, A. D., T. N. Salova, and P. M. Turekhanova. 2003. "The wild fruit and nut plants of Kazakhstan." *Horticultural Reviews* 29: 305–71.

"Explorers find valuable apple germplasm in remote areas of former USSR." *Diversity* (1994). Reprinted in *Seed Savers Exchange* (Summer, 1994): 74–82.

Forsline, Phillip L., Herb S. Aldwinckle, Elizabeth E. Dickson, James J. Lugy, and Stan G. Hokanson. 2003. "Collection, maintenance, characterization, and utilization of wild apples of Central Asia." *Horticultural Reviews* 29: 1–62.

Morgan, Joan, and Alison Richards. 1993. *The Book of Apples*. London: Ebury Press.

Pollan, Michael. 2000. *The Botany of Desire*. New York: Penguin Press.

Pringle, Peter. 2008. *The Murder of Nikolai Vavilov*. New York: Simon & Schuster.

Ratcliffe, I. C., and G. M. Henebry. 2005. "Urban land cover change analysis: The value of comparing historical spy photos with contemporary digital imagery." *International Archives of the Photogrammetry, Remote Sensing and Spatial Information Sciences* 36(8): 1682–77.

Rosenzweig, Michael L. 2003. *Win-Win Ecology: How the Earth's Species Can Survive in the Midst of Human Enterprise*. New York: Oxford University Press.

Vavilov, Nikolay Ivanovich. 1929. *The Role of Central Asia in the Origin of Cultivated Plants*. Leningrad: Bureau of Applied Botany.

———. 1931. "The wild relatives of fruit trees of the Asian part of the USSR and Caucasus, and the problem of the origin of fruit trees." *Transactions in Applied Botany, Genetics and Breeding* 26(3).

———. 1997. *Five Continents*. Edited by L. E. Rodin, translated from the Russian by Doris Love. Rome: IPGRI (Bioversity International); St. Petersburg: VIR.

Vitkovich, Victor. 1960. *Kirghizia Today: Travel Notes*. Translated by David Skvirsky. Moscow: Foreign Language Publications.

Chapter Nine: Rediscovering America and Surviving the Dust Bowl: The U.S. Southwest

Breshears, David D., et al. 2005. "Regional vegetation die-off in response to global-change-type drought." *Proceedings of the National Academy of Sciences* 102(42): 15144–48.

Calloway, Doris H., R. D. Giaque, and F. M. Costa. 1974. "The superior mineral content of some American Indian foods in comparison to federally donated counterpart commodities." *Ecology of Food and Nutrition* 3: 203–11.

Fryer, E. R., ed. 1941. *Statistical Summary, Human Dependency Survey, Navajo Reservation.* Windowrock, AZ: Division of Socio-Economic Surveys, Soil Conservation Service, Navajo Area.

Nabhan, Gary Paul, and Shawn Kelly. 2005. "Changes in Hopi agrobiodiversity, 1893–2005." Oral presentation at Uto-Aztecan Conference, Northern Arizona University, Flagstaff.

Livingston, Matt, Debra, Moon, Cornelia Butler-Flora, and Maria E. Fernandez. 2002. *Hopi Community Food Project.* Second Mesa, AZ: Hopi Pu'tavi Office.

Nabhan, Gary. 2004. "Bringing back desert springs." *Yes!* (Winter). www.yesmagazine.org/article.asp?ID=693.

Nabhan, G. P., C. W. Weber, and J. W. Berry. 1985. "Variation in the composition of Hopi Indian beans." *Ecology of Food and Nutrition* 16(2): 135–52.

Pringle, Peter. 2008. *The Murder of Nikolai Vavilov.* New York: Simon & Schuster.

Soil Conservation Service. 1939. *Statistical Summary, Human Dependency Survey, Navajo and Hopi Reservations.* Windowrock, AZ: Division of Socio-Economic Surveys, Section of Conservation Economics, Soil Conservation Service, Navajo Area.

Soleri, Daniela, and David A. Cleveland. 1993. "Hopi crop diversity and change." *Journal of Ethnobiology* 13: 203–31.

Whiting, Alfred F. 1936. "Hopi Indian agriculture II: Seed source and distribution." *Museum of Northern Arizona Museum Notes* 10(5): 13–16.

———. 1942. *Ethnobotany of the Hopi.* Museum of Northern Arizona Press, Flagstaff.

Chapter Ten: Logged Forests and Lost Seeds: The Sierra Madre

Blancas, Leslie. 2001. "Hybridization between rare and common plant relatives: Implications for plant conservation genetics." PhD dissertation, University of California, Riverside.

Bukasov, Sergei M. 1926. "Un hibrido de maiz y *Euchlaena mexicana*." *México Forestal* 38. Reprinted in *Revista Argentina de Agronomia* 5: 113–15.

———. 1931. "The cultivated plants of Mexico, Guatemala and Colombia." Supplement to the *Bulletin of Applied Botany* (Leningrad).

Burns, Barney T., Mahina Drees, Gary P. Nabhan, and Suzanne C. Nelson. 2000. "Crop diversity among indigenous farming cultures in the tropical deciduous forest." In *The Tropical Deciduous Forest of Alamos: Biodiversity of a Threatened Ecosystem in Mexico*, edited by Robert H. Robichaux and David A. Yetman, pp. 152–71. Tucson: University of Arizona Press.

Bye, Robert, Jr. 1994. "Prominence of the Sierra Madre Occidental in the biological diversity of Mexico." In *Biodiversity and Management of the Madrean Archipelago and the Sky Islands of the Southwestern United States and Northwestern Mexico*, edited by L. F. DeBano, P. F. Folliott, A. Ortega Rubios, G. J. Gottfried, R. H. Hamre, and C. B. Edminster, pp. 19–27. General Technical Report RM-GTR-264. Fort Collins, CO: U.S. Department of Agriculture.

CECCAM 2003. *Boletin de prensa colectivo de comunidades indigenas y campesinas de Oaxaca, Puebla, Chihuahua, Veracruz. Mexico.* Press release, October 9.

Cohen, Barry Mendel. 1980. "Nikolai Ivanovich Vavilov:His Life and Work." PhD. dissertation, University of Texas, Austin.

Doebley, John, and Gary Paul Nabhan. 1989. "Further evidence regarding gene flow between maize and teosinte." *Maize Genetics Cooperative Newsletter* 63: 107–109.

Gingrich, Randall W. "The political ecology of deforestation in the Sierra Madre Occidental of Chihuahua." www.sierramadrealliance.org/sierra-pol-ecol/Deforestation.pdf.

Greenpeace. 2006. "Maseca lies to Mexican consumers." GE Food Alert Campaign Center, www.agobservatory.org/headlines.cfm?archive=6_2006.

LaRochelle, Serge, and Fikret Berkes. 2003. "Traditional ecological knowledge and practice for wild edible plants: Biodiversity use by the Raramuri in the Sierra Tarahumara, Mexico." *International Journal for Sustainable Development and World Ecology* 10: 361–75.

Lumholtz, Carl. 1902. *Unknown Mexico.* New York: Scribner's and Sons.

Mares Trías, Albino. 1999. *Comida de los Tarahumaras.* CONACULTA/Culturas Populares, Mexico City, D.F., Mexico. Don Burgess McGuire, trans.

Martinez Juarez, Victor, Horacio Almanza Alcalde, and Augusto Urteaga Castro Pozo. 2006. *Diagnostico Sociocultural de Diez Municipios de la Sierra Tarahumara.* Report for the Sierra Madre Alliance, Chihuahua City, Chihuahua, Mexico.

Mayer, George. 1996. *Sobre los conflictos sociales, económicos, ecológicos e interétnicos en la Sierra Tarahumara del estado de Chihuahua.* Informe para la Secretaria de Relaciones Exteriores de los Estados Unidos Mexicanos, Chihuahua City, Chihuahua, Mexico.

Nabhan, Gary Paul. 1989. *Enduring Seeds: Native American Agriculture and Wild Plant Conservation.* San Francisco: North Point Press.

Nabhan, Gary Paul, Barney T. Burns, Mahina Drees, Robert Bye Jr., Peter Warshall, Diana Hadley, and Richard Felger. 1994. "North Sierra Madre Occidental and its Apachean outliers: A neglected center of biodiversity." *In Biodiversity and the Management of the Madrean Archipelago and the Sky Islands of the Southwestern United States and Northwestern Mexico,* edited by L. F. De Bano, R. H. Hamre. pp. 36–59. General Technical Report RM-GTR-264. Fort Collins, CO: U.S. Department of Agriculture.

Pennington, Campbell W. 1963. *The Tarahumar of Mexico: Their Environment and Material Culture.* Salt Lake City: University of Utah Press.

———. 1969. *The Tepehuan of Chihuahua: Their Material Culture.* Salt Lake City: University of Utah Press.

Sanchez-Gonzalez, José de Jesús, and José Ariel Ruz Corral. 1997. "Teosinte distribution in Mexico." *In Gene Flow among Maize Land Races, Improved Maize Varieties, and Teosinte: Implications for Transgenic Maize,* edited by J. Antonio Serratos, Martha C. Willcox, and Fernando Castillo Gonzalez, pp. 20–39. El Batan, Mexico: CIMMYT.

Vavilov, Nikolay Ivanovich. 1925. November 17 letter to Sergey Mikhaylovich [Busakov] in Mexico City. Reprinted in Vavilov, Nikolay Ivanovich, 1931, "Mexico and Central America as the principal centre for the origin of cultivated plants in the New World." *Bulletin of Applied Botany, Genetics and Plant Breeding* 36: 1–179 (Leningrad).

———. 1931. "Mexico and Central America as the principal centre for the origin of cultivated plants in the New World." *Bulletin of Applied Botany, Genetics and Plant Breeding* 36: 1–179 (Leningrad).

Wilkes, H. Garrison. 1965. "Teosinte introgression in the maize of the Nobogame Valley." *Harvard University Botanical Museum Leaflets* 22: 297–311.

———. 1997. "Teosinte in Mexico: Personal retrospective and assessment." *In Gene Flow among Maize Land Races, Improved Maize Varieties, and Teosinte: Implications for Transgenic Maize,* edited by J. Antonio Serratos, Martha C. Willcox, and Fernando Castillo Gonzalez, pp. 1–8. El Batan, Mexico: CIMMYT.

Chapter Eleven: Deep into the Tropical Forests of the Amazon

Balee, William, and A. Gely. 1989. "Managed forest succession in Amazonia: The Kápor case." In *Resource Management in Amazonia: Indigenous and Folk Strategies,* edited by Darrell A. Posey and William Balee. Advances in Economic Botany Series, number 7. New York: New York Botanical Garden.

Caqueta moist forests (NTO1070). 2007. World Wildlife Fund Terrestrial Ecosystems. www.worldwildlife.org/wildworld/profiles/terrestrial/nt/nt0107_full.htm.

Cohen, Barry Mendel. 1980. "Nicolai Ivanovich Vavilov: His life and work." PhD dissertation, University of Texas, Austin.

———. 1991. "Nicolai Ivanovich Vavilov: The explorer and plant collector." *Economic Botany* 45(1): 38–46.

Correal, Camilo, Germán Zuluaga, Liliana Madrigal, Sonia Calcedo, Antonia Mutumbajoy, Lida Garces, Diva Devia, Eva Yela Bernards. and Delsy Burgos. 2007. *Ingano, Colombia: Community Food System Data Tables.* Montreal, Canada: Center for Indigenous People's Nutrition and Environment, McGill University.

Giraldo, Iganacio, Iván Sarmiento, Fabio Quevedo, Silvia Amaya, and Germán Zuluaga. 2004. *The Ethnobiological Study of the Liana Paullinia Yoco, Indicator of the Status of Biological and Cultural Conservation of the Amazonian Piedmont.* Arlington, VA: Amazon Conservation Team.

Hawkes, Jack G. 1999. "The evidence for the extent of N. I. Vavilov's New World Andean centers of cultivated plant origins." *Genetic Resources and Crop Evolution* 46: 163–68.

Hecht, Susanna B. 1990. "Indigenous soil management in the Latin American Tropics: Neglected knowledge of native people." In *Agroecology and Small Farm Development*, edited by Miguel A. Altieri and Susanna B. Hecht. Boca Raton, FL: CRC Press.

Kravchenko, Victor. 1946. *I Chose Freedom: The Personal and Political Life of a Soviet Official.* New York: Scribner.

Posey, Darrell. 1985. "Indigenous management of tropical forest ecosystems: The case of the Kayapó Indians of the Brazilian Amazon." *Agroforestry Systems* 3: 139–58.

Pringle, Peter. 2008. *The Murder of Nikolai Vavilov*. New York: Simon & Schuster.

Sarmiento Combariza, Iván, and Beatriz Alzáte Atehortúa. 2005. *The Spatial Analysis of the Transformation of the Forests of the Colombian Amazonian Piedmont*. Arlington, VA: Amazon Conservation Team.

Schultes, Richard Evans. 1994. "The importance of ethnobotany in environmental conservation." *American Journal of Economics and Sociology* 53(2): 202–206.

Vavilov, Nikolay Ivanovich. 1939. "The important agricultural crops of pre-Colombian America and their mutual relationships." *Publications of the National Department of Geography* 71(10): 1–55.

———. 1997. *Five Continents*. Edited by L. E. Rodin, translated by Doris Love. Rome: IPGPRI (Bioversity); St. Petersburg: VIR.

Chapter Twelve: The Last Expedition

Cohen, Barry Mendel. 1980. "Nicolai Ivanovich Vavilov: His life and work." PhD dissertation, University of Texas, Austin.

Golubev, G. 1987. *Nikolai Vavilov: The Great Sower—Pages from the Life of the Scientist.* Moscow: Mir Publishers. Originally published in Russian in 1979, translated into English by Vadim Sternik.

Popovsky, Mark. 1984. *The Vavilov Affair*. North Haven, CT: Archon Books.

Vavilov, Nikolay Ivanovich. 1997. *Five Continents*. Edited by L. E. Rodin, translated by Doris Love. Rome: IPGPRI (Bioversity); St. Petersburg: VIR.

Epilogue

Fowler, Cary. 2006. "Mud, blood and genes." *Crop Diversity Topics: Analysis and Reflections*, no. 5. Global Crop Diversity Trust. www.croptrust.org/documents/newsletter/newsletter_croptrust_v5_final.htm.

Hamilton, Neil. In press. *Food Democracy and the Future of American Values*. Des Moines, IA: Drake University Agricultural Law Center. http://students.law.drake.edu/ agLaw Journal/pastIssues/Vol9No1/hamilton.html.

Lappé, Frances Moore, Joseph Collins, and Peter Rosset. 1998. *World Hunger: Twelve Myths*, 2nd ed. New York: Grove/Atlantic.

Mushita, Andrew, and Carol B. Thompson. 2007. *Biopiracy of Biodiversity: Global Exchange as Enclosure*. Trenton, NJ: Africa World Press.

Pearce, Fred. 2005. "Return to Eden." *New Scientist* (January 22).

Pringle, Peter. 2008. *The Murder of Nikolai Vavilov*. New York: Simon & Schuster.

Russell, Sharman Apt. 2005. *Hunger: An Unnatural History*. New York: Basic Books.

Seabrook, John. 2007. "Sowing for the apocalypse: The quest for a global seed bank." *The New Yorker* (August 7): 60–71.

Vavilov, Nikolay Ivanovich. 1997. *Five Continents*. Edited by L. E. Rodin, translated from the Russian by Doris Love. Rome: IPGRI (Bioversity); St. Petersburg: VIR.

Acknowledgments

My first thanks is to my dear friend Dr. Ken Wilson, executive director of the Christensen Fund, who proposed this book project to me, guided it along its way, and generously supported it through the foundation in innumerable ways. My second thanks is to my wife, Dr. Laurie Monti, who not only accompanied me on most of the trips recounted in this book, but also added her own skills as ethnographer, oral history archivist, and ethnomusicologist to the mix. Other friends joined us on some of the trips and offered their own insights: Kent Whealy, Sergey Alexanian, David Cavagnaro, Dave Denny, Mark Plotkin, Liliana Madrigal, Barney Burns, Shawn Kelly, Walt Coward, Karim-Aly Kassam, Rafique Keshavjee, Randy Gingrich, Kamal Mouzawak, Rami Zuryak, and Dalia Al-Jawhary, to name a few. All added their own unique insights to this project.

I am especially indebted to the farmers, herders, market vendors, and agricultural scientists of the many countries I visited while retracing the routes of Nikolay Vavilov. I can only name a few of the many here, but they include Ogonazar Aknazarov, Fathi Malim, Ahmed Jerry Hammam, Omar Ahmed Ali, Fred Kabotie, Ferrell Secakuku, Richard Pentewa, Dr. Melaku Worede, Dr. Tewolde Birhan Gebre Egziabher, Dr. Sue Edwards, Dr. Aimak Dzangaliev, Dr. Tatiana Salova, Dr. Sulaiman Al-Khanjari, Dr. Germán Zuluaga, and Mahdi el-Hewety. I am deeply grateful to Nikolay Vavilov's son, Dr. Yuri Vavilov, for speaking with me while I was in Russia, and to the many others who have written gracefully about Vavilov: Mark Popovsky, Frank Browning, Cary Fowler, Pat Mooney, Peter Pringle, and G. A. Golubev.

For permission to quote Vavilov's field journals and use his photos, I thank the Director General of VIR, Nikolay Dzyubenko, as well as Dr. Sergey Alexanian of VIR, and Director General Emile Frison of Bioversity International. Jonathan Cobb and Chuck Savitt at Island Press have been hugely committed to this project and brought wisdom as well as good taste to it. Editors Jennifer Sahn and Chip Blake at *Orion* magazine are also among the literary support team. Thanks also to Jeremy Cherfas and Tim Tracy for web-based follow-up.

Finally, I wish to acknowledge the support team at the Center for Sustainable Environments, which helped me through the logistic and graphic exercises of these journeys: Jeanette Sherman, Heather Farley, DeJa Walker, Paige Irwin, Dan Boone, Joan Carstensen, and Patty West. I also wish to acknowledge my former housemate in the Stinkin' Hot Desert—Dr. William Feldman—who first sang the words "Nikolay Vavilov" to me in 1977. What a long, strange trip it has been.

No plants were harmed or collected in the making of this book.

About the Author

Gary Paul Nabhan is a world-renowned conservationist, ethnobiologist, agricultural historian, and essayist. Nabhan is the author of *Why Some Like It Hot* (Island Press, 2006), *Coming Home to Eat*, and many other books and articles printed in five languages. His collaborative conservation work has been honored with a MacArthur "Genius" Fellowship and the Lifetime Achievement Award from the Society for Conservation Biology and the Quivira Coalition. His creative nonfiction has received a Lannan Literary Award and The John Burroughs Medal for nature writing. Founder and facilitator of the Renewing America's Food Traditions collaborative, he is currently a Research Social Scientist at the Southwest Center at the University of Arizona. See www.garynabhan.com to track his lecture and photo exhibit schedules.

Index

About Island Press

Since 1984, the nonprofit Island Press has been stimulating, shaping, and communicating the ideas that are essential for solving environmental problems worldwide. With more than 800 titles in print and some 40 new releases each year, we are the nation's leading publisher on environmental issues. We identify innovative thinkers and emerging trends in the environmental field. We work with world-renowned experts and authors to develop cross-disciplinary solutions to environmental challenges.

Island Press designs and implements coordinated book publication campaigns in order to communicate our critical messages in print, in person, and online using the latest technologies, programs, and the media. Our goal: to reach targeted audiences—scientists, policymakers, environmental advocates, the media, and concerned citizens—who can and will take action to protect the plants and animals that enrich our world, the ecosystems we need to survive, the water we drink, and the air we breathe.

Island Press gratefully acknowledges the support of its work by the Agua Fund, Inc., Annenberg Foundation, The Christensen Fund, The Nathan Cummings Foundation, The Geraldine R. Dodge Foundation, Doris Duke Charitable Foundation, The Educational Foundation of America, Betsy and Jesse Fink Foundation, The William and Flora Hewlett Foundation, The Kendeda Fund, The Forrest and Frances Lattner Foundation, The Andrew W. Mellon Foundation, The Curtis and Edith Munson Foundation, Oak Foundation, The Overbrook Foundation, the David and Lucile Packard Foundation, The Summit Fund of Washington, Trust for Architectural Easements, Wallace Global Fund, The Winslow Foundation, and other generous donors.